JN087821

英和対照［工学基礎テキスト］シリーズ

English–Japanese Bilingual Textbook Series of Fundamental Engineering

熱 力 学
Thermodynamics

新井雅隆／古畑朋彦

Masataka Arai and Tomohiko Furuhata

森北出版株式会社
Morikita Publishing Co., Ltd.

Published by
Morikita Publishing Co., Ltd.

English-Japanese Bilingual Textbook Series of Fundamental Engineering Thermodynamics

●本書のサポート情報を当社 Web サイトに掲載する場合があります．下記の URL にアクセスし，サポートの案内をご覧ください．

https://www.morikita.co.jp/support/

●本書の内容に関するご質問は下記のメールアドレスまでお願いします．なお，電話でのご質問には応じかねますので，あらかじめご了承ください．

editor@morikita.co.jp

●本書により得られた情報の使用から生じるいかなる損害についても，当社および本書の著者は責任を負わないものとします．

英和対照「工学基礎テキスト」シリーズの刊行に寄せて
Preface for the publication of English-Japanese bilingual textbook series in fundamental engineering

People say that the best way to learn a foreign language is to marry a native speaker. While this is not practical for most people, it means that the important thing is to "live" the language.

This textbook series is designed for foreign students who wish to learn not only engineering subjects in Japan, but also the Japanese language and Japanese business customs in a short time. It is expected that this series can also be useful for Japanese students who wish to improve their English language skills.

This series focuses on the field of mechanical engineering since "Monodukuri" is a common fundamental subject for all students of engineering. The series deals only with the minimum requirements of each subject necessary in order to understand the content in both Japanese and English. For advanced levels of each subject, please refer to higher level Japanese or English textbooks on these fields.

In this project entitled "Education program for advanced and top-level production leaders in Asian countries", Asian students invited by MEXT will take professional courses specially developed by Gunma University in cooperation with consortium companies. After completing the "production leadership course", the students are expected to be hired by the consortium companies or their daughter companies in Asian countries.

The first edition of this textbook series for foreign students of Gunma University was published by the "Advanced education program for career development of foreign students from Asia at the Asian human resource fund" supported by METI: Ministry of Economy, Trade

昔から，外国語をマスターするには，目的とする言語を母国語とする人と結婚することが一番であるといわれています．しかし，実際にこれを実行することは現実的ではなく，本当の意味は「目的とする外国語の中で生活することが秘訣である」と理解されます．

このテキストシリーズは，日本で工学の専門分野を学び，さらに日本語と日本のビジネス習慣を短期間で修得しようとする外国人留学生を対象に企画されました．しかし，英語力を高めようとする日本人学生にも役立つものと期待しています．

本シリーズの各テキストの内容は，機械工学を中心としています．これは，工学部に所属するすべての学生にとって「ものづくり」が工学部の共通基礎科目であると考えたためです．しかし，日本語を学ぶのに最低限必要な専門分野のみを扱っていますので，より高度な内容はそれぞれの専門教科書を参照してください．

本学で実施している「先進・高度ものづくりリーダーの育成プロジェクト」は，アジア諸国から文部科学省国費留学生を募集し，コンソーシアム企業のご協力のもと，特別プログラムを提供しており，修了後はコンソーシアム企業および地域の日本企業に就職し，日本とアジア諸国をつなぐものづくりリーダーを育成することを目的としています．

本テキストシリーズの初版は，経済産業省と文部科学省の「アジア人財資金構想」高度専門留学生育成事業の支援を受け，本プロジェクトに所属する留学生向けのテキストとして制作されました．

and Industry and MEXT: Ministry of Education, Culture, Sports, Science and Technology.

However, numerous comments were received suggesting the usefulness of this textbook series for all students of engineering in Japan including foreign students. The textbooks were then revised and published by Morikita Publishing Co., Ltd.

We hope the torchbearers of the next generation will find this series helpful. Lastly, we appreciate the contribution of the parties involved in this project.

Publication committee of the Asian Human Resource Project of Gunma University. Chairman: Tomio Obokata, Professor Emeritus

　しかし，初版を印刷の後，本シリーズは留学生を含めて日本の工学系の学生にも役立つので，市販してはどうかと多くのご意見をいただきました．その結果，再編集のうえ，このたび森北出版より刊行されることになりました．

　これからの時代を担う方々のお役に立つものと期待し，また，出版にご協力いただいた関係の皆様に感謝し序とします．

群馬大学アジア人財出版委員会
委員長　小保方富夫（名誉教授）

Preface

Thermodynamics is one of the fundamental subjects of dynamics for scientists and mechanical engineers. The classical thermodynamics that had been developed during 18th century with the development of various kinds of heat engines is still fundamentals for all thermal scientific and engineering fields. The first and the second law of thermodynamics are the fundamental laws not only for the energy science but also for national and social sciences for human being.

Deep understanding on the thermodynamics is then needed for all the scientists and engineers to establish their own professional achievement. Many scientists and engineers in various countries had been contributing to establish the thermodynamics as a professional subject of dynamics. Then the thermodynamics had been written down with many languages. Also it has been translated to other ones and worldwide standard for thermodynamics has been established.

To learn and to teach the professional subject such as thermodynamics, text written by native language is convenient. However there are another conveniences in a study of national science using the second or third language. Bilingual study on a professional subject of the science will bring us many merits. For example, a fundamental concept expressed with different languages will be received more deep understanding.

This bilingual text is edited for this purpose and readers can find different styles of expressions to explain a unique thermodynamic concept that has scientific nature independent from languages. For persons who will firstly lean the thermodynamics, a professional text written their native language is recommended to avoid miss understandings for fundamental terms. After leaning fundamentals, this text gives them a

序　文

　熱力学は，科学者や機械技術者にとって基礎的な力学の領域の一つである．18世紀におけるさまざまな熱機関の開発とともに発展してきた古典熱力学は，熱に関するすべての科学と工学の領域にとって，依然としていまでも，その基礎となっている．熱力学の第一および第二法則は，熱に関する科学だけでなく，自然科学や人類のための社会科学についての基本法則である．

　それゆえ，すべての科学者や技術者が自らの専門的知識を高めて完成させるためには，熱力学についての深い理解が必要である．力学の一専門領域として熱力学が完成されたが，そのためには異なる国々の多くの科学者や技術者の貢献が必要であった．その結果として，熱力学は多くの言語で記述されている．さらにまたそれは他の言語に翻訳され，熱力学のついての普遍的な内容の精選が行われてきた．

　熱力学のような専門領域を学びかつ教えるためには，母国語で書かれた専門書が便利である．しかし，第二または第三言語を使って自然科学を学ぶことには別の都合のよいこともある．自然科学の専門領域についての2か国語学習には多くのメリットがある．たとえば，一つの基本概念が二つの異なる言語で表現されているので，その概念はより深く理解されるようになる．

　この2か国語の本書は，このような深い理解を得るために編集されたものであり，言語自体とは無関係な科学的特性をもつ固有な基礎概念を説明するための異なった表現形式を読者は見いだすことが可能になる．熱力学を初めて学ぼうとする人達には，その人達の固有の言語で記述してある熱力学の専門書を，基礎的な内容における誤解を防ぐためにお勧めする．基礎を学んだあと，本書はさらに深い理解を得るためにたいへん役に立つことに

great contribution for deeper understanding. Also it is useful for Japanese readers to lean various ways of English expressions, and for other readers to lean Japanese expressions.

The contents of this text are limited within a part of the classical thermodynamic concerning energy balance, entropy change and efficiency of cycles, because of the high priority of them for the scientific and technical importance. The text is composed with 15 chapters corresponding with 15 lessons of 30 hours. It means that it is edited as a sub-text of university class level, and we believe that readers can lean both thermodynamics and professional expressions of Japanese or English.

Autumn 2008

なる．さらに日本人の読者が英語の種々の表現を学ぶためにこれは有効であり，また他国の読者が日本語の表現を学ぶためにも有効である．

本書の内容は，古典熱力学の一部であり，科学技術における重要性が高いことから，内容はエネルギーバランス，エントロピー変化，サイクルの効率に限定されている．本書は，15 回 30 時間の学習に対応した 15 章から成り立っている．このことは大学の講義水準の補助専門書として編集してあることを示していて，読者が熱力学はもちろんであるが，それだけでなく日本語や英語による専門的な表現を学ぶことに役立つと考えている．

2008 年秋

English-Japanese Bilingual Textbook Series of Fundamental Engineering
Thermodynamics
Authors
Masataka ARAI and Tomohiko FURUHATA

英和対照「工学基礎テキスト」シリーズ
熱力学

著者
新井雅隆，古畑朋彦

Contents
目　次

Nomenclatures 用 語

A : Area 面積 m^2

C_s : Sound Velocity 音速 m/s

c : Specific Heat 比熱 J/(kg·K)

c_p : Specific Heat at Constant Pressure 定圧比熱 J/(kg·K)

c_v : Specific Heat at Constant Volume 定容比熱 J/(kg·K)

E : Energy エネルギー J

F : Force 力 N

f : Equation of State 状態方程式

G : Mass 質量 kg

$\dot{G}$: Mass Flow Rate 質量流量 kg/s

g : Acceleration of Gravity 重力加速度 $\mathrm{m/s}^2$

H : Enthalpy エンタルピー J

h : Specific Enthalpy 比エンタルピー J/kg

h : Height 高さ m

L : Technical Work 工業仕事 J/kg

L^G : Technical Work 工業仕事 J

M : Molecular Weight 分子量 kg/mol

m : Mass of Substance 物質の質量 kg

n : Polytropic Exponent ポリとロープ指数

P : Pressure 圧力 Pa

Q : Quantity of Heat 熱量 J

q : Quantity of Heat 熱量 J/kg

R : Gas Constant 気体定数 J/(kg·K)

R_0 : Universal Gas Constant 一般気体定数 J/(K·mol)

r : Latent Heat for Vaporization 蒸発潜熱 J/kg

S : Entropy エントロピー J/K

s : Specific Entropy 比エントロピー J/(kg·K)

s : Distance 距離 m

T : Temperature 温度 K

t : Time 時間 s

U : Internal Energy 内部エネルギー J

u : Specific Internal Energy 比内部エネルギー J/kg

V : Volume 容積 / 体積 m^3

v : Specific Volume 比容積 / 比体積 m^3/kg

W : Expansion Work 膨張仕事 J/kg

W^G : Expansion Work 膨張仕事 J

w : Velocity 速度 m/s

x : Quality of Dryness 乾き度

x : Coordinate 座標 m

z : Vertical Coordinate 垂直座標 m

ε : Compression Ratio 圧縮比

η : Thermal Efficiency 熱効率

κ : Ratio of the Specific Heat 比熱比

Chapter 1
Concept of Thermodynamics
熱力学の概念

1.1 Heat Engines and Thermodynamics

Heat engine is a kind of energy conversion system in which mechanical work is performed. Figure 1-1 shows an example of a heat engine. In the early days, it was a steam plant used to draw up water.

Heat from combustion is transferred to a boiler and its phenomena are treated as a heat transfer problem in the engineering field. High pressure steam in the boiler is introduced into the cylinder. It acts on the piston and mechanical work is produced by piston movement. This process is called the thermal expansion process of steam. With the aid of mechanics and hydrodynamics engineerings, the task of drawing wa-

1.1 熱機関と熱力学

熱機関は，機械的な仕事を行うエネルギー変換システムの一つである．図1-1に熱機関の例を示す．これは，初期の蒸気プラントであり，水を汲み上げるために使われていたものである．

燃焼によって得られた熱はボイラに伝えられるが，この現象は工学においては熱伝達問題として扱われている．ボイラの中の高圧蒸気はシリンダに送られる．蒸気はピストンに作用し，ピストンの動きにより機械的な仕事が実現される．このプロセスは，蒸気の熱膨張プロセスとよばれるものである．機構学および流体力学という工学技術が関与して水の汲み上げが行われる．シリンダ-ピストンの

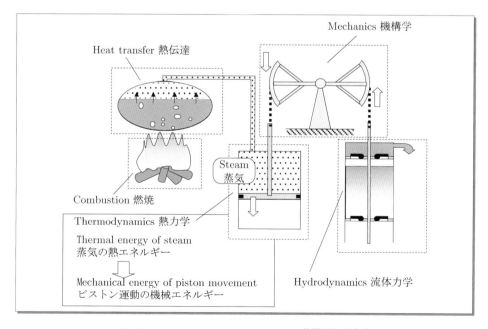

Fig.1-1 Heat engine and thermodynamics 熱機関と熱力学

ter is accomplished. Steam expansion in a cylinder-piston system is one of the key phenomena of the heat engine and its engineering is called thermodynamics.

From the end of the 17th century, steam was recognized as a power source for machines and by the 18th century the steam engine became the prime mover of water pumps and other machines. The engineers' aim to produce high power and raise efficiency spurred the development of prime movers. Thermodynamics was developed in the 18th-19th centuries with the aid of theoretical considerations of heat engines such as the steam engine.

Classical thermodynamics is a macroscopic view of thermodynamics and gives the dynamic relationship between heat and mechanical work. Even up to the present days, it is significant to the development of high efficiency heat engines. Energy conversion is a field of engineering thermodynamics; molecular dynamics and chemical reactions concerning heat is a field of chemical thermodynamics. Statistical thermodynamics, which was one of the origins of quantum physics, became the foundation of engineering and chemical thermodynamics. Science and engineering fields being covered with thermodynamics are illustrated in Fig.1-2.

システムにおける蒸気の膨張は，ここに例として示した熱機関の基本的な事象であり，このような工業技術にかかわる工学を熱力学とよんでいる．

17世紀の末から，蒸気は機械の動力源として認識され，18世紀には蒸気機関は水汲みポンプやその他の機械の原動機として発達してきた．技術者は，高出力で効率の高い原動機を目指して開発を行ってきた．熱力学は，蒸気機関のような熱機関についての理論的な検討の促進とともに，18世紀から19世紀にかけて確立された学問体系である．

古典熱力学は巨視的な熱力学であり，熱と機械的な仕事を結びつける力学的関係を与えている．現在においても，古典熱力学は高効率機関の開発にとって必要な学問になっている．エネルギー変換は工業熱力学の分野であり，また，分子動力学や熱が関与する化学反応は化学熱力学の分野である．統計熱力学は量子物理学の一つの出発点であり，工業熱力学や化学熱力学の基盤として発展してきたものである．熱力学の範疇にある科学技術の領域を図1-2に示す．

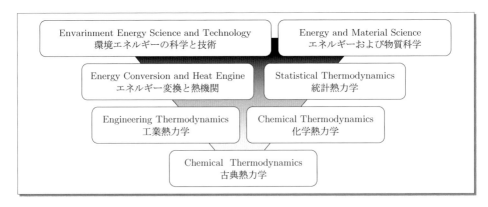

Fig.1-2   Classification of thermodynamics   熱力学の分類

1.2 Temperature and Heat

Thermodynamic conditions of material especially gaseous material can be specified with pressure, density, and temperature. Pressure and density are sense perceived quantities and are easily measured.

However, temperature measurement had been difficult in the old days. The general meanings of temperature and heat are different but these have not been separated in the early days of thermodynamics. The distinction between the meanings of temperature and heat was only attained after the development of the thermometer.

As for temperature measurement, three kinds of scale are used. Fahrenheit temperature [°F] was developed based on the temperature scale of the environment. Celsius temperature [°C] was established with the development in steam engineering; Kelvin temperature (thermodynamic temperature, absolute temperature) [K] was introduced with the development of classical thermodynamics.

Scale conversions among the three temperature scales mentioned above are as follows:

1.2 温度と熱

物質，とくに気体状態の物質の熱力学的な状況は，圧力，密度，温度によって特定することができる．圧力と密度は感覚的にとらえられる量であり，計測することも比較的簡単である．

一方，温度の計測は，以前はたいへんむずかしいものであった．温度と熱の普遍的な意味は異なるが，熱力学が確立する以前においては両者の区分がされていなかった．温度と熱を区分して扱うことは温度計が開発されたあとに初めて可能となった．

温度の計測では，3 種類の温度目盛が使われている．華氏温度［°F］といわれる温度は，環境の温度（日常的な気温）を基準としたスケールの目盛りである．摂氏温度［°C］は，蒸気機関などの工業技術とともに確立され，また，ケルビン温度（熱力学温度，絶対温度）［K］は，古典熱力学の確立の際に導入された温度である．

これらの 3 種類の温度目盛の間の変換は，以下の式を用いて行うことができる．

$$T[\mathrm{K}] = 273.15 + T^C[^\circ\mathrm{C}] = 273.15 + \frac{5}{9}\left(T^F[^\circ\mathrm{F}] - 32\right) \tag{1-1}$$

$$T^C[^\circ\mathrm{C}] = T[\mathrm{K}] - 273.15 = \frac{5}{9}\left(T^F[^\circ\mathrm{F}] - 32\right) \tag{1-2}$$

$$T^F[^\circ\mathrm{F}] = \frac{9}{5}T^C[^\circ\mathrm{C}] + 32 = \frac{9}{5}\left(T[\mathrm{K}] - 273.15\right) + 32 \tag{1-3}$$

Table 1-1 shows the international temperature scale and presents a list of temperatures that are usually experienced in human life. It is clear that the Fahrenheit temperature is a good index of environmental temperature in daily life.

Heat, which is different from temperature, is a kind of energy that is spontaneously moving from one object with a high temperature to other object with a low temperature. The quantity of heat 1 [J] is defined as the heat which is equiva-

表 1-1 は，国際実用温度目盛と，われわれの生活において体験するいくつかの温度を示したものである．生活の場における環境温度に対して都合のよい指標が華氏温度であることは，この表から明らかである．

熱はエネルギーの一種である．熱は温度の高い物体から温度の低い他の物体に移動するものであり，温度とは異なる．熱量 1 ［J］は 1 ［N］の力によって 1 ［m］の移動仕事によって定義されている．

Table 1-1   Temperature scale   温度目盛

International temperature scale 国際実用温度目盛（IPTS-68）			
Equilibrium state 平衡状態	T_{68} [K]	T^C [°C]	T^F [°F]
Triple point of equilibrium hydrogen 平衡水素の三重点	13.81	− 259.34	− 434.81
Boiling point of equilibrium hydrogen at 25/76 atm 平衡水素の 25/76 気圧の沸点	17.042	− 256.108	− 428.99
Boiling point of equilibrium hydrogen 平衡水素の沸点	20.28	− 252.87	− 423.17
Boiling point of neon ネオンの沸点	27.102	− 246.048	− 410.89
Triple point of oxygen 酸素の三重点	54.361	− 218.789	− 361.82
Boiling point of oxygen 酸素の沸点	90.188	− 182.962	− 297.33
Triple point of water 水の三重点	273.16	0.01	32.02
Boiling point of water 水の沸点	373.15	100	212.00
Solidifying point of tin 錫（すず）の凝固点	505.1181	231.9681	449.54
Solidifying point of zinc 亜鉛（あえん）の凝固点	692.73	419.58	787.24
Solidifying point of silver 銀の凝固点	1235.08	961.93	1763.47
Solidifying point of gold 金の凝固点	1337.58	1064.43	1947.97
Guideline of temperatures 温度の目安			
Typical temperature index 代表的温度指標	T_{68} [K]	T^C [°C]	T^F [°F]
Severe cold 極寒	255	− 18	0
Cold 寒冷	268	− 5	23
Icing point of water 水の氷点	273	0	32
Chilly 冷気	278	5	41
Comfortable temperature 快適温度	293	20	68
Body temperature 体温（人体）	309	36	97
High fever temperature 高熱（人体）	313	40	104
Boiling point of water 水の沸点	373	100	212

lent to 1 [m] movement work done by a force of 1 [N].

$$1\,[\mathrm{J}] = 1\,[\mathrm{N\cdot m}] = 2.388459 \times 10^{-1}\,[\mathrm{cal}] \tag{1-4}$$

Furthermore, the quantity of heat 1 [cal] is defined as the heat which can result from a temperature rise of 1 [°C] in 1 [g] mass of water at 15 [°C], and this can be converted to a Joule unit of heat using the next equation.

さらに，熱量 1［cal］は 15［℃］の水 1［g］が 1［℃］上昇する温度として定められ，その熱量は，つぎの式によりジュール単位の熱に換算することができる．

$$1\,[\mathrm{cal}] = 4.1868\,[\mathrm{J}] \tag{1-5}$$

It should be noted that the Joule unit is independent of temperature scale while the calorific unit of heat is subjected to the definition of the temperature of Celsius scale. The differences between those two definitions should be clearly understood.

ここで，ジュール単位は温度目盛とは無関係であるが，カロリー単位の熱は温度の摂氏目盛の温度の定義を基礎とし，それに依存した状態となっている．この二つの熱量定義の相違を理解することはたいへん重要である．

1.3 Thermodynamic Quantity

1.3 熱力学的諸量

As mentioned before, thermodynamic conditions of objects can be specified with pressure, density, and temperature. Hence, these quantities are called thermodynamic quantities of state. Internal energy, entropy, etc. which are introduced in classical thermodynamics are also quantities of thermodynamic state.

前述したように，対象物の熱力学的な状況は圧力，密度，温度によって特定することができる．したがって，圧力，密度，温度などは熱力学的状態量とよばれている．古典熱力学で導入する内部エネルギーやエントロピーなども，熱力学的状態量である．

On the other hand, heat moves from one object with a high temperature to other object with low temperature and is not an inherent property of object. Thus, heat is called the non-state quantity (variable) of thermodynamics. Work is also a non-state quantity (variable) since it is independent from the object that is recieved the work. Further, it is completely converted to heat.

一方，熱は温度の高い物体から温度の低い物体に移動し，熱自体は物体の本質的な特性ではない．そこで熱は熱力学的非状態量（変量）とよばれている．仕事は，仕事を受ける物体とは無関係であるので，仕事もまた熱力学的非状態量（変量）である．さらに，仕事は熱に完全に変換される．

Thermodynamic quantities of state and non-state variables are classified into two groups. One group is the extensive quantity of state and the other is the intensive quantity of state. Mass, volume, heat, etc. can be divided by quantity of the object and are called extensive quantities. While temperature, pressure, etc. cannot be divided and thus, keep the same quantities even when the object is divided. These are called intensive quantities.

熱力学的状態量や非状態変量は，二つのグループに分けることができる．一つのグループは，示量性量とよばれるものであり，他のグループは，示強性量とよばれるものである．質量，体積，熱などは，対象物とともに分割することが可能であるので，示量性量とよばれている．反対に，温度や圧力などは，対象物を分割した場合にも分けることができず，同じ量を維持するので，示強性量とよばれている．

1.4 Thermodynamic Equilibrium and Quasi-Static Change

1.4 熱力学的平衡と準静的変化

A substance under specified thermodynamic condition and the ambient surroundings around it is called a thermodynamic system. Thermodynamic equilibrium state of the system means that intensive quantities of thermodynamic state are in equilibrium in both the substance and the surroundings as illustrated in Fig.1-3. It means

熱力学的に対象としている物体とその周囲環境を合わせて熱力学的システムとよんでいる．システムの熱力学的平衡状態とは，対象物体と外界において示強性量である熱力学的状態量が平衡状態にあることをいう．また，その状態は図1-3に示してある．この状態は，温度差と圧力差が対象物体と外界の間に存在

that there are no temperature and pressure differences between the substance and the surroundings. No heat transfer results between them and the system is said to be in thermal equilibrium.

When the substance is processed from state 1 to state 2, the temperature and pressure of the substance would be changed with the process. It usually means that temperature and pressure differences between the substance and the surroundings would result and thus, the state of thermal equilibrium could not be maintained in the process. In other words, the substance during this process is isolated from the surroundings.

There is another kind of processes in which the substance is not isolated and can not change its thermodynamic state with no change of the surroundings. Here, the following assumptions are introduced in the process. The change pro-

しないことを意味している．さらにその結果として，両者のあいだに熱の移動がないことから，熱平衡とよばれている．

いま，物体が状態 1 から状態 2 に変化したとすれば，物体の温度や圧力はこの状態変化とともに変化する．このことは，物体と外界の間に温度差や圧力差が生じる結果となり，熱平衡状態は状態変化の途中では維持されないことになる．言い換えると，この変化過程においては，物体は外界とは無関係に孤立しているといえる．

一方，物体は孤立しておらず，外界の変化をともなうことなしに，その熱力学的状態を変えられないという別の変化過程も存在する．ここで，つぎのような仮定を変化過程に導入する．変化が少しずつ進行するもの

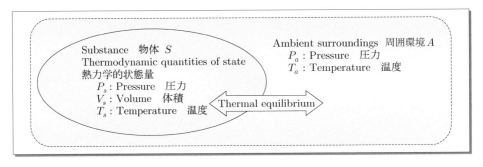

Fig.1-3　Thermal equilibrium state　熱力学的平衡状態

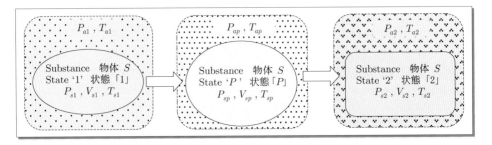

Fig.1-4　Quasi-static process　準静的変化

ceeds very slowly and the surrounding thermo-dynamic quantities such as temperature and pressure are spontaneously changed to keep the same values of the substance in the process as shown in Fig.1-4.

In this process, the thermal equilibrium state would be completely kept throughout the pro-cess. This is an ideal process that can never be realized. However, this process is simple and has lots of merits in making theoretical ap-proaches. The process near this change is called a quasi-static process because slow change in the process is similar to the ideal thermodynamic equilibrium process.

In the field of classical thermodynamics, mac-roscopic phenomena of thermodynamics are usually considered and analyzed using the basic assumption of a quasi-static process. Thus, the concept of a quasi-static process is the basis of thermodynamics.

とし，外界の温度や圧力などの外界の熱力学的状態量が，変化の途中において図1-4に示すように，物体と同じ値をとるように，対象物体の変化に追従して変化していくものとする.

この変化過程では，その全過程において熱平衡の状態が完全に維持されることになる. この過程は仮想的な過程であり，実現することはできないが，単純な過程となり，理論的なアプローチを行うための多くのメリットを有している. ゆっくりとした変化は仮想的な熱力学的平衡変化過程に近いものとなるので，平衡状態を近似的に維持する変化を準静的変化とよんでいる.

古典熱力学の分野では，熱力学上の巨視的な現象をこの準静的変化の仮定のもとに検討し，解析する. したがって，準静的変化の概念は熱力学の基礎と認識すべきである.

1.5 Pressure balance for gas filled in a container

1.5 容器に充填されたガスの圧力バランス

To express the thermodynamic change of the gaseous substance, we often use a simple gas state in a container. A container in vacant state as shown with Fig.1-5 (a) means that no gas is filled in it. When a gas is filled in the container, pressures of the gas and the ambient surround-ings are balanced ($P_g = P_a$) as indicated in Fig.1-5 (b). Here, piston with no mass and no friction for its movement are assumed.

When a weight of mass m is added on the piston (Fig.1-5 (c)), pressure of gas should be expressed as follows.

気体物体の熱力学的変化を表現する場合，容器内の単純なガスの状態がしばしば使われる. 図1-5 (a) は真空状態の容器を示したもので，容器内に気体がまったく充填されていないことを意味している. 気体が容器内に充填されていれば，気体の圧力と周囲環境の圧力は図1-5 (b) のようにつり合って（$P_g = P_a$）いる. なお，質量が無視できるピストンで，かつその動きに対して摩擦がないことをここでは想定している.

質量 m のおもりがピストンに作用している図1-5 (c) の場合では，気体の圧力は以下の式で表現される.

$$P_g = P_a + \frac{mg}{\text{Area of piston}} \tag{1-6}$$

It means that the pressure of the gas in the container is balanced with the total pressure of ambient surroundings and additional weight. The pressure P_g is sometimes called as an abso-

このことは，容器内の気体の圧力が，周囲環境の圧力と追加されたおもりの総合圧力とつり合っていることを示している. 圧力 P_g は気体の絶対圧とよばれることがあり，また

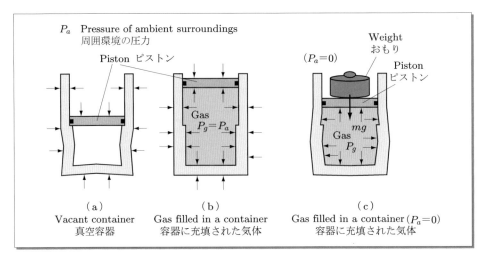

Fig.1-5 Pressure balance for gas filled in a container
容器に充填された気体の圧力バランス

lute pressure of gas, and $\Delta P = P_g - P_a$ is called as a gauge pressure because of it's easy measurement using pressure gauge.

If no consideration is taken for ambient pressure, in other words, a system composed with weight, piston, container, and gaseous substance in it is assumed, the pressure of ambient surroundings should be omitted. This situation seems that the system is placed in the surroundings with zero pressure and the gas pressure can be simply expressed as follows.

$\Delta P = P_g - P_a$ は圧力ゲージを用いて簡単に測定できることからゲージ圧とよばれる.

もしここで，周囲の圧力を考えないならば，言い換えれば，おもり，ピストン，容器　および容器内の気体物体から構成されている系を想定するならば，周囲環境の圧力は省略すべきである．この状況は圧力ゼロの外界の中に想定している系が置かれている状況であり，この場合では気体の圧力はつぎのように簡単に表現することができる.

$$P_a = 0 \quad \Rightarrow \quad P_g = \frac{mg}{\text{Area of piston}} \tag{1-7}$$

This treatment is often used as a simple pressure balance in the thermodynamic analysis explained in the following chapters. Cylinder-piston system such as Fig.2-2 in the next chapter is a typical one of this treatment and Eq.1-7 and Eq.2-10 have the same meaning.

この取り扱いは，以後の章で説明する熱力学的解析における単純な圧力のつり合いとして，しばしば用いられている．次章の図2-2のようなシリンダ-ピストン系はこの取り扱いの典型的なものであり，式 (2-10) と式 (1-7) は同じ内容である.

<table><tr><td>**Problems**</td><td>**問題**</td></tr></table>

△1-1

(1) Convert 25 °C to Kelvin temperature.

△1-1

(1) 25 °C をケルビン温度に変換しなさい.

(2) Convert 86 °F to Celsius temperature.

△1-2

Convert 120 kgf/cm^2 to Pa unit.

△1-3

Convert 150 kcal to Joule unit.

△1-4

Are specific quantities such as specific volume v, specific enthalpy h, intensive or extensive?

（2）86 °F を摂氏温度に変換しなさい.

△1-2

120 kgf/cm^2 を Pa 単位に変換しなさい.

△1-3

150 kcal をジュール単位に変換しなさい.

△1-4

比体積や比エントロピーのような単位質量あたりの量は示強性であるか, 示量性であるか.

Chapter 2
Work and Heat
仕事と熱

2.1 Energy and Work

Energy is the general concept of the quantity that has a potential of activities. Potential energy concerning gravitational force, kinetic energy owing to a velocity and thermal energy concerning temperature are three major energy states in the field of thermodynamics. Also work is a kind of energy and energy conversion from thermal energy to work is one of the main topics of the thermodynamics. Using the definitions of energies listed in Eq.2-1~Eq.2-5, the law of energy conservation is expressed by Eq.2-6.

2.1 エネルギーと仕事

エネルギーは，活動を生みだすもとになるポテンシャル量の一般的概念である．重力に起因する位置エネルギー，速度から求められる運動エネルギー，温度に関連する熱エネルギーが，熱力学分野の主要な3種類のエネルギー状態である．仕事もまたエネルギーの一種であり，熱エネルギーから仕事へのエネルギー変換は，熱力学の主要分野の一つである．式（2-1）～（2-5）に掲げた式を用いれば，エネルギー保存則は式（2-6）のように書き表すことができる．

Potential energy　位置エネルギー $$E_p = mgz \tag{2-1}$$

Kinetic energy　運動エネルギー $$E_k = \frac{1}{2}mw^2 \tag{2-2}$$

Thermal energy　熱エネルギー $$E_{th} = mcT \tag{2-3}$$

Other kinds of energy　その他のエネルギー $$E_{others} \tag{2-4}$$

Work　仕事 $$W \tag{2-5}$$

$$E_{total} = \sum (E_p + E_k + E_{th} + E_{others} + W) = \text{constant} \tag{2-6}$$

Definition of work is simply expressed by Fig2-1(a) and Eq.(2-7). When the force is acted on the different direction of a movement as shown Fig2-1(b), work can be obtained inner product between force and movement vectors as

仕事は図2-1（a）と式（2-7）により簡単に表現することができる．力が図2-1（b）のように移動方向と異なる向きに作用する場合では，力と移動方向のベクトルの内積により，式（2-8）から求めることができる．

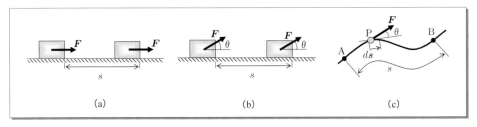

Fig.2-1　Work　仕事

Eq.2-8.

Figure 2-1(c) shows the general definition of work done by the force F. Then the general equation of work can be expressed as shown in Eq.2-9.

図 2-1（c）は力 F によって行われる仕事の一般的な定義である．この図より仕事の一般的な表現は式（2-9）のように表すことができる．

$$W = (\boldsymbol{F} \cdot \boldsymbol{s}) = Fs \tag{2-7}$$

$$W = (\boldsymbol{F} \cdot \boldsymbol{s}) = Fs \cdot \cos\theta \tag{2-8}$$

$$W = \int_A^B \boldsymbol{F} \cdot d\boldsymbol{s} = \int_A^B F\cos\theta \cdot ds \tag{2-9}$$

2.2 Work by Thermal Expansion of Gaseous Substance

2.2 気体の膨張による仕事

Cylinder-piston system shown in Fig.2-2 is considered as a simple thermodynamic system for analysis. Bottom of the cylinder is heated with a flame and heat is transferred from the flame to a contained gaseous substance in the cylinder so as to expand the gaseous substance. Piston can be moved smoothly without friction and weight of mass m is loaded on the piston to balance the gaseous pressure in the cylinder.

The initial state of gaseous substance is noted by subscript 1 and the final state is shown by subscript 3. Other symbols for analysis are followed to the illustrations in the figure. Pressure

図 2-2 のようなシリンダ-ピストン系を，解析のための簡単な熱力学的システムとして考える．シリンダの底は，ほのおにより加熱され，ほのおからの熱はシリンダ内の気体状物体に伝えられる．また，その結果として気体状物体は膨張する．ピストンは摩擦なしに滑らかに動くことができ，質量 m のおもりが気体状物体の圧力とつり合うようにピストンに載せられている．

気体状物体の初期状態を状態1とし，その最終状態を状態3とする．解析のための他の変数などは，図中に示してある表現のとおりとする．また，状態1から状態3までの変化

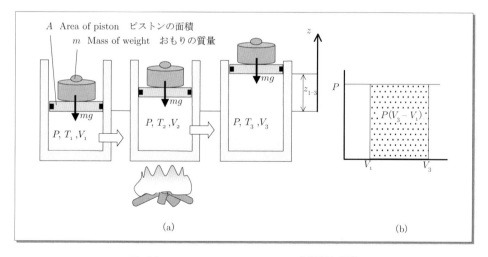

Fig.2-2　Thermal expansion and work　熱膨張と仕事

diagram during the change from state 1 to state 3 is illustrated in the P-V chart as shown in Fig.2-2(b).

During an expansion process of the gaseous substance, it acts to lift the piston against the load of the piston. Its lift force is produced with the pressure of the substance in the cylinder. Then the lift force can be expressed by the mass of weight on the piston.

$$F = PA = mg \tag{2-10}$$

The gaseous substance work done to the piston is obtained with the force and movement of piston. And it can be expressed using the displacement volume of the piston.

$$W = \int_1^3 F dz = \int_1^3 PA dz = PAz = P(V_3 - V_1) \tag{2-11}$$

On the other hand, a potential energy of the weight increases during the upward movement caused the expansion of the gaseous substance in the cylinder. According to the conservation law of energy, the following relationship can be obtained for constant pressure expansion of the gaseous substance.

$$W = \int_1^3 F dz = \int_1^3 mg dz = mgz \tag{2-12}$$

$$W = mgz = P(V_3 - V_1) \tag{2-13}$$

2.3 Weight Work to Spring

Consider the weight work to a spring. When a weight of mass m is hung slowly and statically to a spring, spring is expanded as shown in Fig.2-3(a).

According the Hooke's law, spring force F is proportional to an elongation of the spring and is balanced to the gravitational force of weight. The spring factor k can be obtained with the balance of spring force and gravitational force of weigh. The work received by the spring can be calculated with a F-x diagram of Fig.2-3(b) and Eq.2-15.

の間の圧力は，図 2-2（b）に示した P-V 線図のようになる．

気体状物体の膨張過程において，気体は荷重に抗する向きにピストンを持ち上げる．この持ち上げ力は，シリンダ内の物体の圧力によって生じているので，持ち上げ力は，ピストンの上に置かれたおもりの質量を用いて表すことが可能である．

ピストンに与える気体状物体の仕事は上記の力 F とピストンの動きから求めることができる．さらにそれは，ピストンの行程容積を用いて表すことが可能になる．

一方，シリンダ内の気体状物体の膨張にともなってピストンとおもりが上方に移動する間に，おもりの位置エネルギーは増加する．エネルギーの保存則に従えば，気体状物体の定圧膨張に対して，つぎのような関係が得られる．

2.3 ばねに対するおもりの仕事

おもりがばねにする仕事を考える．質量 m のおもりを，ゆっくりかつ準静的にばねに吊るすと，ばねは図 2-3（a）のように伸びる．

フックの法則によれば，ばねの力 F は，ばねの伸びに比例していて，おもりの自重による力とつり合っている．ばね定数 k は，ばねの力とおもりの自重のつり合いから求められる．ばねが得た仕事は図 2-3（b）の F-x 線図と式（2-15）から求めることができる．

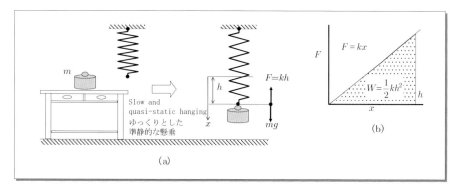

Fig.2-3　Quasi-static work to spring　ばねに対する準静的仕事

$$h = \frac{mg}{k} \tag{2-14}$$

$$W_{spring} = \int_0^h Fdx = \int_0^h kxdx = \frac{1}{2}kh^2 \tag{2-15}$$

In this case, dynamic energy of weight motion is neglected and all process is supposed to be done slowly and statically. Then it is a kind of quasi-static process of weight hanging.

When the weight is suddenly released after connecting to the spring of free condition, as shown by Fig.2-4, it would start moving down and up so as to show a vibration, and its movement should not be neglected.

In this case, equation of motion for weight

この場合は，おもりの運動による運動エネルギーは無視されていて，すべての変化はゆっくりと準静的に行われているとしている．したがって，これは準静的な操作としておもりを懸垂したことになる．

自由状態のばねにおもりを取り付けたあと，図 2-4 に示すように急に落下させたとすると，おもりは落下と上昇を行い振動状態となる．したがって，おもりの動きを無視できないことになる．

この場合，おもりの運動方程式はつぎのよ

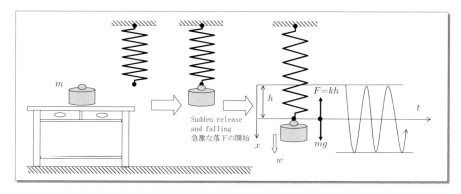

Fig.2-4　Dynamic work to spring　ばねに対する動的仕事

became the following.

うになる.

$$mg = kx + mw\frac{dw}{dx} \qquad (2\text{-}16)$$

It can be easily integrated and we can understand that potential energy of weight is converted to the spring energy and dynamic energy of weight as shown by Eq.2-18.

これは積分が可能であり，結果として，おもりの位置エネルギーはばねのエネルギーとおもりの運動エネルギーに変換されていることが，式（2-18）より明らかである.

$$W_{weight} = \int_0^x mg dx = \int_0^x kx dx + \frac{1}{2}mw^2 \qquad (2\text{-}17)$$

$$mgx = \frac{1}{2}kx^2 + \frac{1}{2}mw^2 \qquad (2\text{-}18)$$

The maximum down position of weight appears at the position indicated by Eq.2-19 where the velocity of the weight becomes zero. This position is twice down position indicated by Eq.2-14. The potential energy loss of weight at this position during weight falling from start to the maximum position is balance to the work received by the weight because dynamic energy of weight at this position is zero.

The maximum velocity appears at the static balance position of Eq.2-14 and the velocity is indicated by Eq.2-22.

おもりの最大降下位置は，式（2-19）で示される位置として表され，そこではおもりの速度がゼロとなる.また，その位置は，式(2-14)に示した位置の2倍のところとなる.その位置では，おもりの運動エネルギーがゼロであるから，最初の状態から最大位置までの落下でおもりが失った位置エネルギーは，ばねになされた仕事とバランスしている.

おもりの最大速度は，式（2-14）に示した静的なつり合い位置で現れ，その速度は式（2-22）となる.

$$h_{max} = \frac{2mg}{k} \qquad \text{at} \quad w = 0 \qquad (2\text{-}19)$$

$$E_{spring/max} = \frac{1}{2}kh_{max}^2 = \frac{1}{2} \cdot \frac{2mg}{h_{max}}h_{max}^2 = mgh_{max} \qquad (2\text{-}20)$$

$$x_{w \cdot max} = \frac{mg}{k} \qquad \text{at} \quad \frac{dw}{dx} = 0 \qquad (2\text{-}21)$$

$$w_{max} = \sqrt{\frac{m}{k}}g \qquad \text{at} \quad h = \frac{mg}{k} \qquad (2\text{-}22)$$

2.4 Thermal Energy

2.4 熱エネルギー

Thermal energy is usually expressed by a production of heat capacity and temperature. While, quantity of heat is defined with a product of heat capacity and temperature difference. As shown in Fig.2-5, substantial quantity of heat is

熱エネルギーは，通常，熱容量と温度の積で表されている．一方，熱量は熱容量と温度差で示されている．図2-5に示すように，熱量の本質は温度に無関係であり，熱量は高温の物体から低温の物体に移動する．温度とは

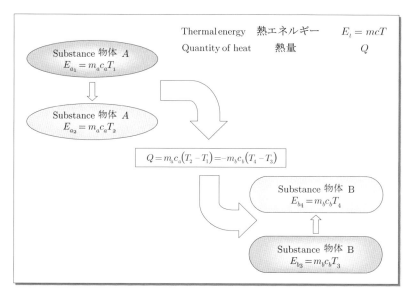

Fig.2-5　Thermal energy and quantity of heat　熱エネルギーと熱量

independent of temperature and it is transferred from a high temperature material to a low temperature material. Quantity of heat that is independently defined from temperature is more convenient when it is compared with other kinds of energy.

Here, let's consider the energy conversion from thermal energy to kinetic energy and potential energy. According to the illustration of Fig.2-6, thermal energy of 2.1 kJ is supposed to be assigned to a glass of water of which mass is 0.1 kg. When this energy is assigned as a quantity of heat, it can elevate the water temperature by 5 degrees. Whereas, if it is assigned by velocity increase, it is equivalent of the kinetic energy of 205 m/s and is equivalent with the potential energy of 2143 m.

無関係に定義されている熱量であるならば, 熱量を別の種類のエネルギーと比較する場合にさらに好都合になる.

ここで, 熱エネルギーから他のエネルギーへの変換を検討してみよう. まず, 図 2-6 のイラストに従い, コップに入っている質量 0.1 kg の水に 2.1 kJ の熱エネルギーを賦与することを想定する. このエネルギーが熱量として賦与される場合では, このエネルギーにより水の温度は 5℃上昇する. ところが, もしこのエネルギーが速度増加として与えられるものとすれば, それは 205 m/s の速度の運動エネルギーに相当し, それは 2 143 m の高さの位置エネルギーに相当する.

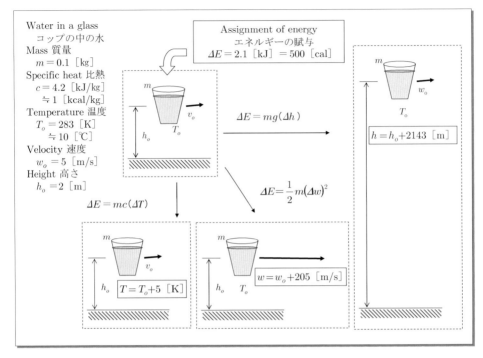

Fig.2-6   Assignment of energy   エネルギーの賦与

Problems

△2-1

A friction less piston and cylinder device (see Fig.2-7) initially contains 1 m³ of air at 0.2 MPa. Now the air is heated and the volume of air increases to 4 m³. During this process the pressure remains constant. Determine the work done by the air.

△2-2

1 kg of water is dropped from a height of 1000 m. If the potential energy of water is fully converted into heat and the heat is fully used to raise the temperature of water, determine the temperature increase ΔT of the water. Here, the specific heat of water is 4.185 kJ/(kg·K) and the gravitational acceleration is 9.807 m/s².

問題

△2-1

摩擦のないピストンとシリンダからなる装置 (図 2-7) の内部に，初期状態で 0.2 MPa の空気が 1 m³ 入っている．その空気を加熱したところ，その体積が 4 m³ に増加した．この増加の過程で圧力は一定のままであった．このとき空気がした仕事を求めなさい．

△2-2

1 kg の水が 1 000 m の高さから落下する．もし位置エネルギーがすべて熱に変化し，その熱のすべてが水の温度を上昇するために用いられたとするとき，その水の温度上昇 ΔT を求めなさい．ここで，水の比熱は 4.185 kJ/(kg·K)，重力加速度は 9.807 m/s² とする．

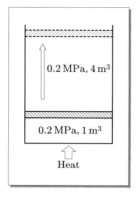

Fig.2-7　Schematic for Problem 2-1　問題 2-1 の説明図

Chapter 3
Nature of Gas and Ideal Gas
気体の性質と理想気体

3.1 States of Substance

Solid, liquid, and gas are the three major states of a substance. Substance in a solid state keeps its configuration. Substance in a liquid state can change its configuration and has a free surface when it is pored into a vessel with an opening. It can easily flow out through a hole at the bottom of the vessel. Gas can expand in every direction and can change its volume to fit the capacity of a vessel.

Generally, a substance at low temperature is in a solid state and it changes to liquid or to gas with temperature increase. The state of the substance is also affected by its pressure that is usually equivalent to the pressure of the surroundings. Gaseous state can be changed to liquid or solid states by elevating the pressure, however, pressure is less effective than temperature.

Here, we consider the gaseous state of a substance in a vessel as shown in Fig.3-1. Mole-

3.1 物体の状態

固体と液体，および気体は，物体の主要な3種類の状態である．固体状態の物体は，物体自身が示す形状を保っている．液体状態の物体は，物体自身の形状を変えることができ，開口部のある容器に注ぎ込めば自由界面をもつことができる．また，容器の底の穴から簡単に流出することもある．気体状態の物体はすべての方向に膨張することができ，容器の容積に合わせ，容器を満たすように気体自身の体積を変えることができる．

一般的に，低温の物体は固体状態であり，温度の増加につれ，液体または気体に変化する．物体の状態は，物体自身の圧力，一般的には物体の圧力に等しい周囲の圧力によっても変化する．圧力を高めることにより，気体の状態であったものを液体または固体に変えることができるが，その作用は，温度の場合ほど顕著なものではない．

ここで，図 3-1 に示したような容器内の物体の気体状態について検討してみよう．物体

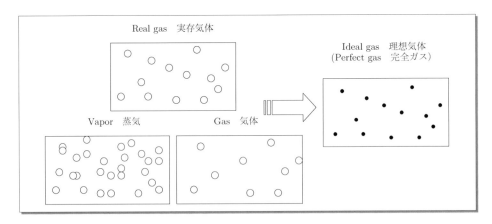

Fig.3-1 Gaseous substance 気体状態の物体

cules of the substance, which has their own volumes, move around everywhere in the vessel. Steam is a gas state where its temperature and pressure are not far from the state of liquid; however, there is no substantial difference between gas and steam. To simplify the theoretical treatment of thermodynamics, an ideal gas (perfect gas) that does not have its own molecular volume is introduced. Using the concept of a perfect gas, we can treat the general molecular behavior of many kinds of gaseous substances.

States of a substance such as solid, liquid, and gas can be usually expressed on a pressure-temperature (P-T) diagram. Figure 3-2 shows examples of P-T phase diagrams. Brief explanations of the triple point and critical point are shown as unique points on the diagram.

The triple point of carbon dioxide appears at 526.9 kPa and –56.4 °C where the solid, liquid, and gas states of carbon dioxide co-exist. Since the pressure of the critical point is so high, the density of a gaseous substance near the critical point is not so different from that of liquid. A

の分子は，物体特有の体積をもっていて，かつ容器内のいたるところを動き回っている．蒸気は気体状態の一つであり，その温度や圧力が物体の液体状態に近い場合であるが，蒸気と気体の間には本質的な差違は存在していない．熱力学における理論的な扱いを単純化するために，分子自身の体積がないものとした理想気体（完全気体）を導入する．この完全気体の概念を用いることにより，気体状態の多くの物体について気体分子の一般的な挙動を論じることができるようになる．

固体，液体，気体という物体の状態は，一般的には圧力と温度の線図（P-T線図）上で表現することができる．図3-2はP-T相図の例と，相図上の特異点として現れる三重点と臨界点の簡単な説明を行ったものである．

二酸化炭素の三重点は526.9 kPa，－56.4 ℃のところに現れ，そこでは固体，液体，気体状態の二酸化炭素が共存している．臨界点における圧力はたいへん高いので，臨界点付近の気体状態の物体の密度と，液体状態の密度の差違は小さくなる．温度と圧力が臨界温度

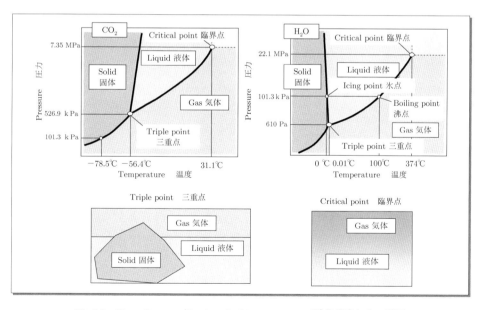

Fig.3-2 Phase diagrams of carbon dioxide and water 　二酸化炭素と水の相図

substance whose state is over the critical temperature and pressure shows similarities to the liquid and gas states. Thus, this state should be considered as the fourth state of the substance.

Water is a popular substance. The boiling point of water is 101.3 kPa and 100 °C, which co-exists in both liquid and gas. The icing point of water is 101.3 kPa and 0 °C, This means that the co-existing point of solid and liquid is different from the triple point at 0.61 kPa and 0.1 °C. A most unique characteristic which is the inverse effect of pressure on the boundary between the solid and liquid states can be recognized in the P-T diagram of water. In other words, an increase in pressure leads to the melting of the solid water (ice).

と臨界圧力を超える物体の状態は，一部は液体に，また別の部分は気体に似た状態になっている．そこで，このような状態は物体の第4の状態と考えるべきである．

水はどこにでもある物体である．水の沸点は 101.3 kPa，100 °C であり，そこでは液体と気体の水が共存している．水の氷点は 101.3 kPa，0 °C であり，これは固体と液体の水の共存を意味している．またこの点は，0.61 kPa，0.1 °C である水の三重点とは異なっている．水の P-T 相図上でもっとも変わった特徴が固体相から液体相への境界における圧力の逆向き効果として示されている．すなわち，圧力の増加が固体（氷）の融解をもたらすということである．

3.2 Equation of State

3.2 状態方程式

The thermodynamic state of a substance can be uniquely defined with the physical quantities such as pressure, volume, and temperature. Thus, these quantities are called the fundamental quantities of thermodynamic state.

物体の熱力学的状態は，圧力，体積，温度のような物理量によって一義的に定めることができる．したがって，このような物理量は熱力学的基本的状態量とよばれている．

P : Pressure 圧力
V : Volume 体積
T : Temperature 温度
Fundamental quantities of thermodynamic state
熱力学的基本状態量

As seen in Fig.3-2, the thermodynamic state can be expressed on the P-T phase diagram which means that there is a functional relationship among P, V and T.

The equation of state has various formes of expressions with the same meaning that define the thermodynamic state of a substance. Eq.3-1 shows the equation of state. When the thermodynamic state changes from state 1 to state 2 (Fig.3-3), two sets of thermodynamic quantities such as (P_1, V_1, T_1) and (P_2, V_2, T_2) are obtained but both should satisfy the same equation of state.

図 3-2 に示したように，物体の熱力学的状態は P-T 相図で表すことができ，このことは，P, V, T の間に一つの関数関係が存在していることを意味している．

状態方程式は，種々の表現形式をもつが，いずれの式も物体の熱力学的状態を規定するという同じ意味をもっている．式 (3-1) は，この状態方程式の例である．熱力学的状態が状態 1 から状態 2 に変化する場合では（図 3-3），(P_1, V_1, T_1) と (P_2, V_2, T_2) という二組みの状態量が得られるが，その両者とも同じ状態方程式を満足することになる．

$$f(P, V, T) = 0 \quad \begin{cases} P = f_{V,T}(V, T) \\ V = f_{T,P}(T, P) \\ T = f_{P,V}(P, V) \end{cases} \quad (3\text{-}1)$$

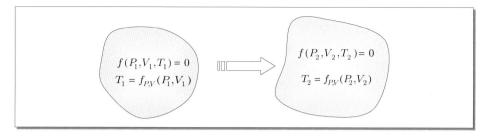

Fig.3-3　Equation of state and thermodynamic fundamental quantities of state
状態方程式と熱力学的基本状態量

As for an ideal gas (perfect gas), the equation of state is simply written using Eq.3-2 and Eq.3-3

理想気体（完全ガス）の状態方程式については，式（3-2）や式（3-3）を用いて簡単に書き表すことができる．

$$PV = GRT \tag{3-2}$$
$$Pv = RT \tag{3-3}$$

Where, G is the mass and v is the specific volume of a gaseous substance. R is called the

ここで，G は気体物体の質量，また v はその比容積である．R はガス定数とよばれている．

Table 3-1　Equation of state and related physical constants for gas
理想気体の状態方程式と関連する物理定数

Gaseous volume of 1 mole molecule　　1 モルの分子の気体としての体積

$V(1\,\mathrm{mol\ at\ }273.15\,\mathrm{K\ and\ }1.01325\times10^5\,\mathrm{Pa}\,(760\,\mathrm{mmHg}))$
$= 22.4140\times10^{-3}\,[\mathrm{m}^3]$

Avogadro's number　アボガドロ数	Boltzmann's constant　ボルツマン定数
$N_A = 6.022045\times10^{23}\,[1/\mathrm{mol}]$	$k_\mathrm{B} = 1.380662\times10^{-23}\,[\mathrm{J/K}]$

Universal gas constant　一般ガス定数
$R_0 = N_A k_\mathrm{B} = 8.31447\,[\mathrm{J/(mol\cdot K)}]$

Gas constant　ガス定数

$R = \dfrac{R_0}{M} = \dfrac{8.31441\times10^3}{M}\,[\mathrm{J/(kg\cdot K)}]$ 　　　M : molecular weight（mass）分子量 [kg/kmol]

Equation of state for ideal gas of G [kg] mass　　質量 G [kg] の理想気体の状態方程式
$PV = GRT$

Specific volume　比容積

$v = \dfrac{V}{G}$

Equation of state for ideal gas of 1 kg mass　　質量 1 kg の理想気体の状態方程式
$Pv = RT$

gas constant. It is one of the fundamental physical properties of the gas and is closely related to Avogadro's number and Boltzmann's constant. The fundamental equations and constants concerning the equation of state for an ideal gas are listed in Table 3-1.

これは気体の基本的な物性値の一つであり，アボガドロ数やボルツマン定数と密接な関係をもつものである．理想気体の状態方程式に関連する基本的な式や定数については，表3-1 にまとめてある．

3.3 Specific Heat

3.3 比　熱

Consider the heating process indicated in Fig.3-4. There are two typical heating processes in a gas. One is a constant volume heating process indicated by (a). In this case, the gas pressure increases with heating. The other is a constant pressure heating process indicated by (b) where the gas volume increases with heating. It is known through experience that process (b) needs more quantity of heat than process (a) to attain the same increment of temperature.

図 3-4 に示すような加熱過程を想定する．ガスの加熱過程には二つの典型的な過程がある．その一つは，図（a）に示した定容加熱過程である．この場合では，ガスの圧力は加熱とともに増加する．もう一つの過程は図(b)に示した定圧過程であり，容積は加熱とともに増加する．ここで，温度を同じだけ上昇させるためは図（a）の過程より図（b）の過程のほうが必要な熱量が多いことが経験的に知られている．

Thus, the specific heat which is the heat for a unit temperature increment for a unit mass of gas should be defined for each process, independently.

したがって，単位質量のガスを単位温度だけ上昇させるのに必要な熱量，すなわち比熱については，それぞれの過程ごとに独立して定義すべきであることがわかる．

For process (a), the specific heat at constant volume heating is defined as follows:

図（a）の過程については，定容比熱が以下の式により定義されている．

$$c_v = \frac{\Delta Q_v}{G(T_2 - T_1)} = \frac{\Delta Q_v}{G \cdot \Delta T} = \left(\frac{\partial q}{\partial T}\right)_v \ \left[\text{J/(kg·K)}\right] \quad (3\text{-}4)$$

On the other hand, the specific heat at constant pressure is given by Eq.3-5.

一方，定圧比熱については，式（3-5）で与えられている．

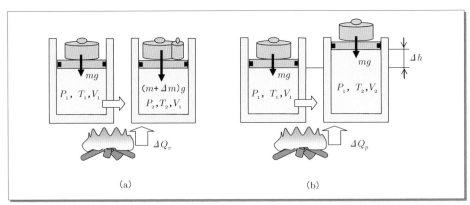

Fig.3-4   Heating process of gas   気体の加熱過程

$$c_p = \frac{\Delta Q_p}{G(T_2 - T_1)} = \frac{\Delta Q_p}{G \cdot \Delta T} = \left(\frac{\partial q}{\partial T}\right)_p \quad \left[\mathrm{J/(kg \cdot K)}\right] \tag{3-5}$$

Using the equation of state for an ideal gas, the quantity of heat that is needed to elevate the temperature of gas in constant pressure heating is noted as follows:

理想気体の状態方程式を使えば，等圧加熱における温度上昇のために必要な熱量は，以下の式のように記述することができる．

$$\Delta Q_p = \Delta Q_v + mg \cdot \Delta h = \Delta Q_v + P_1 \cdot \Delta V = \Delta Q_v + GR \cdot \Delta T \tag{3-6}$$

From Eq.3-6, the following relationship between c_p and c_v can be obtained.

式（3-6）より c_p と c_v の間の以下の関係を求めることができる．

$$c_p = \frac{\Delta Q_p}{G \cdot \Delta T} = \frac{\Delta Q_v + GR \cdot \Delta T}{G \cdot \Delta T} = \frac{\Delta Q_v}{G \cdot \Delta T} + R \tag{3-7}$$

$$c_p = c_v + R \tag{3-8}$$

$$\frac{c_p}{c_v} = \kappa \tag{3-9}$$

Table 3-2　Physical properties of gases　気体の物性値

気体分子	分子量 M		ガス定数 R [J/(kg·K)]	密度 [kg/m³] (0°C, 760 mmHg)	比熱 [kJ/(kg·K)] (0°C, 低圧)		比熱比 κ (c_p/c_v)
	概略値	厳密値			c_p	c_v	
He	4	4.003	2077.2	0.1785	5.238	3.161	1.66
Ar	40	39.944	208.13	1.7834	0.523	0.318	1.66
H_2	2	2.016	4124.4	0.08987	14.248	10.119	1.409
N_2	28	28.016	296.80	1.2505	1.0392	0.7419	1.400
CO	28	28.01	296.83	1.2500	1.0408	0.7432	1.400
NO	30	30.008	277.07	1.3402	0.9981	0.7210	1.385
O_2	32	32.01	259.83	1.42895	0.9144	0.6540	1.399
HCl	36.5	36.465	228.03	1.6391	0.800	0.569	1.40
CO_2	44	44.01	188.92	1.9768	0.8194	0.6301	1.301
N_2O	44	44.016	188.89	1.9878	0.8922	0.7034	1.270
SO_2	64	64.06	129.78	2.9265	0.6083	0.4786	1.272
NH_3	17	17.032	488.20	0.7713	2.056	1.566	1.313
C_2H_2	26	26.036	319.31	1.1709	1.5127	1.2158	1.255
CH_4	16	16.042	518.25	0.7168	2.156	1.633	1.319
C_2H_4	28	28.052	296.37	1.2604	1.612	1.290	1.249
C_2H_6	30	30.068	276.50	1.3560	1.729	1.444	1.20
Air	29	28.964	287.06	1.2928	1.005	0.716	1.402

［佐藤俊，国友孟著：熱力学　第 7 版（1997），26 ページ，丸善株式会社より引用］

$$c_v = \frac{R}{\kappa - 1}, \qquad c_p = \frac{\kappa R}{\kappa - 1} \tag{3-10}$$

The ratio of specific heats defined by Eq.3-9 results from the structure of the molecule and its theoretical value is shown by Eq.3-11. Table 3-2 is a list of the physical properties of real gases.

式 (3-9) によって定義されている比熱比は，分子の構造に依存して得られるものであり，その理論的な値は式 (3-11) に示されている．また，表 3-2 は実際のガスについての物性値である．

$$\left.\begin{array}{l} \kappa(\text{for a single atom molecule}) = \frac{5}{3} = 1.67 \\[2mm] \kappa(\text{for a two atom molecule}) = \frac{7}{5} = 1.40 \\[2mm] \kappa(\text{for a poly} \,(\geq 3)\, \text{atomic molecule}) = \frac{4}{3} = 1.33 \end{array}\right\} \tag{3-11}$$

Further, theoretical derivations of above values are explained in chapter 5.

なお，上記の理論的な値の導出については第 5 章にて説明してある．

3.4 Other State Quantities of Gases

3.4 気体のその他の状態量

Using the specific heat of a constant volume heating process, the internal energy of an ideal gas is defined by Eq.3-12.

等容加熱過程の比熱を使えば，理想気体の内部エネルギーを式 (3-12) により求めることができる．

$$U_i = G\int_0^i c_v dT + U_0 \ [\text{J}] \tag{3-12}$$

U_0 is the base energy at $T = T_0$ and U_i is the internal energy at $T = T_i$. As for a gas of unit mass, the specific internal energy can be noted by u and its total differential du shown in Eq.3-13 is usually used as a point of departure for theoretical treatment of the internal energy.

U_0 は $T = T_0$ に対応した基準状態の内部エネルギーであり，U_i は $T = T_i$ に対応した内部エネルギーである．単位質量の気体については比内部エネルギーがuで記述され，式(3-13) に示されている．その全微分 du が内部エネルギーの理論的な取り扱いの出発点として用いられている．

$$du = c_v dT \ [\text{J/kg}] \tag{3-13}$$

Eq.3-14～3-17 are the definitions of enthalpy H and entropy S of the ideal gas. Similar to the thermodynamic quantities such as pressure, volume, and temperature, internal energy, enthalpy and entropy are also quantities of state. Thus, using the three quantities mentioned above, a thermodynamic process can be expressed as shown in Fig.3-5. The detailed meanings of these thermodynamic quantities are explained

式 (3-14)～(3-17) は，エンタルピー H とエントロピー S の理想気体に対する定義である．圧力，体積，温度などの熱力学的諸量と同様に，内部エネルギー，エンタルピー，エントロピーは状態量である．したがって，ここに示した三つの状態量により，図 3-5 に示したように熱力学的過程を記述することが可能になる．これらの熱力学的状態量の詳細な説明は本テキストの後半にて行う．

later.

$$H_i = G \int_0^i c_p dT + H_0 \ [\mathrm{J}] \tag{3-14}$$

$$dh = c_p dT \ [\mathrm{J/kg}] \tag{3-15}$$

$$S_i = \int_0^i \frac{1}{T} dQ + S_0 \ [\mathrm{J/K}] \tag{3-16}$$

$$ds = \frac{1}{T} dq \ [\mathrm{J/(kg \cdot K)}] \tag{3-17}$$

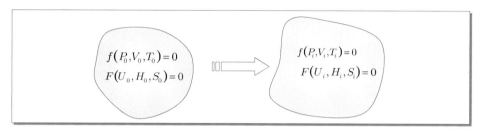

Fig.3-5　Thermodynamic state notated by internal energy, enthalpy and entropy
　　　　内部エネルギー，エンタルピー，エントロピーで記述した熱力学的状態

Problems	問題

△3-1

Determine the gas constant of a real gas whose mass and volume are 1.5 kg and 2 m³ respectively at 200 kPa and 25 ℃.

△3-2

Determine specific volumes of H_2 and CO_2 at 101.3 kPa and 293 K. Those gases are assumed to be ideal gas.

△3-3

In a constant volume heating process, 22.3 kJ of heat is required to heat a gas of 1 kg at 273 K to 303 K. In this case, determine the specific heat for constant volume heating of the gas.

△3-4

Derive Eqs.3-10 from Eqs.3-8 and 3-9.

△3-5

The gas constant and the ratio of specific heats of a gas are 259.83 J/(kg·K) and 1.399, respectively. Determine c_p and c_v of the gas.

△3-1

圧力 200 kPa で温度 25℃ のとき，質量と体積がそれぞれ 1.5 kg，2 m³ である理想気体のガス定数を求めなさい．

△3-2

圧力が 101.3 kPa で温度が 293 K のとき，H_2 と CO_2 の比体積を求めなさい．それらのガスは理想気体であるとする．

△3-3

273 K のガス 1 kg を等容加熱過程において 303 K まで加熱するために 22.3 kJ の熱が必要であった．このとき，このガスの定容比熱を求めなさい．

△3-4

式 (3-8) と式 (3-9) から式 (3-10) を導出しなさい．

△3-5

あるガスのガス定数と比熱比は，それぞれ 259.83 J/(kg·K) と 1.399 である．このガスの c_p と c_v を求めなさい．

△3-6

1 kg of an ideal gas at 300 K is heated at constant pressure to 800 K. In this process, 450 kJ of heat is added to the gas. If the specific heat at constant volume of the gas is 0.75 kJ/(kg·K), determine the gas constant. Moreover, determine the change in internal energy of gas in this process. Here, the specific heat at constant volume is assumed to be independent of temperature.

△3-6

300 K の理想気体 1 kg が一定圧力下で 800 K まで加熱された．この過程でその気体に加えられた熱量は 450 kJ であった．この気体の定容比熱が 0.75 kJ/(kg·K) であるとするとき，気体定数を求めなさい．また，この過程における内部エネルギーの変化量を求めなさい．ここで，定容比熱は温度に依存しないと仮定する．

Chapter 4

The First Law of Thermodynamics
熱力学の第一法則

4.1 Expansion Work

The cylinder-piston system illustrated in Fig.4-1 is a system to consider the work done by an expansion of gas in it. G is the mass of gas whose initial state is noted as state 1. The movement of piston is assumed to be infinitely smooth; no heat loss through the cylinder wall is also assumed to simplify the problem. By the expansion process from state 1 to state 2, the gas pressure is changed as noted in the P-V chart shown in the figure.

According to the definition of work by Eq. 2-7, the work W^G done by the gas of which mass is G [kg] can be noted as follows:

4.1 膨張仕事

図4-1に示したシリンダ-ピストン系は，シリンダ内に充填された気体の膨張による仕事を検討するためのものである．Gは気体の質量であり，その初期状態は状態1と表されている．問題を簡単にするため，ピストンの動きは限りなく滑らかであり，また，シリンダ壁面からの熱損失もないものと想定しておく．状態1から状態2への膨張過程に伴い，気体の圧力は変化するが，この変化は図中のP-V線図上に記載したとおりである．

式（2-7）で定義した仕事によれば，質量G[kg]の気体の行った仕事W^Gは，つぎのように記述することができる．

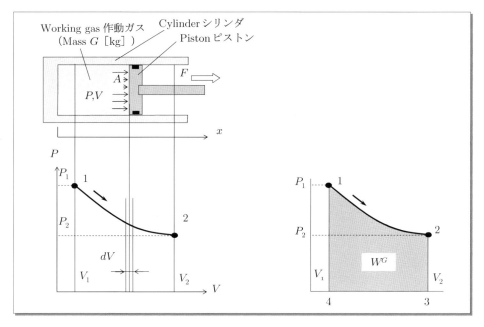

Fig. 4-1　Expansion work done by a gas in a cylinder-piston system
シリンダ-ピストン系内のガスがする膨張仕事

$$W^G = \int_1^2 F dx = \int_1^2 P A dx = \int_1^2 P dV = \text{Area}(1\text{-}2\text{-}3\text{-}4) \tag{4-1}$$

It is clear that the quantity of work corresponds to the Area (1-2-3-4) in the $P\text{-}V$ chart. Using the definition of specific volume, the work can be expressed as follows:

この場合の仕事は，$P\text{-}V$線図上の面積（1-2-3-4）に対応している．比容積の定義を用いると，仕事をつぎのように表現することもできる．

$$v = \frac{V}{G}, \qquad dV = G dv \tag{4-2}$$

$$W^G = G \int_1^2 P dv = G W, \qquad W = \int_1^2 P dv \tag{4-3}$$

Here, W means the work done by a gas of unit mass. Equation 4-4 is a total differential expression of work, and is convenient for theoretical treatment.

ここで，Wは単位質量の気体による仕事である．式（4-4）は仕事の全微分であり，これは理論的な扱いにとって使いやすい表現である．

$$dW = P dv \tag{4-4}$$

Work to the ambient surroundings illustrated in Fig.4-2 should be considered as a general problem of gas expansion work. In this case, gas of state 1 is defined by the state quantities of (P_1, V_1, T_1). These quantities are in an equilibrium condition with the ambient surroundings of (P_a, V_a, T_a).

Through the expansion process, the gas changes to state 2 defined by (P_2, V_2, T_2). During this process, the pressure and temperature of the ambient surroundings are assumed to be simultaneously changed with the state of the gas. In

図4-2に示した外界に対する仕事を，気体の膨張仕事の一般的な問題として取り扱う必要がある．この場合では，状態1の気体は状態量 (P_1, V_1, T_1) で規定されている．また，この状態量は周囲環境の状態量 (P_a, V_a, T_a) と平衡状態になっている．

膨張過程を経て，気体は (P_2, V_2, T_2) で規定される状態2へ変化する．この変化過程の間，周囲環境の圧力と温度は，気体の状態変化に追従して同時に変化していくと想定する．すなわち，気体と周囲環境の間には圧力

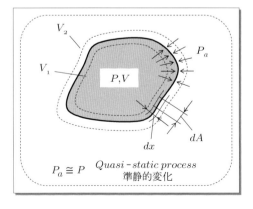

Fig.4-2 Gas expansion work to ambient surroundings
周囲環境に対する気体の膨張仕事

other words, it is assumed that no pressure and no temperature differences between the gas and ambient surroundings exist. Thus, there is no need to include heat loss to the ambient surroundings and unnecessary movements in the ambient surroundings in the problem. This means that a quasi- static process is assumed. The work done by the gas to the ambient surroundings can be expressed by Eq. 4-5.

差や温度差が存在しないと想定する．この結果，気体から外界への熱損失や外界の余分な運動を問題として取り込む必要がなくなる．このような想定が，準静的変化の仮定である．気体が周囲環境に行った仕事は式（4-5）で表すことができる．

$$W^G = \int_1^2 \left(\oint_A PdA \right) dx = \int_1^2 PdV \tag{4-5}$$

Expansion work depends not only on the final expansion state of gas but also on the route of the expansion process itself. Here, consider the different expansion routes illustrated in Fig.4-3. Gas pressures during the expansion processes are noted as P_a, P_b, P_c (Eq. 4-6) corresponding to each expansion route Y_{route}. From Equation 4-1, we can easily derive the comparison among these expansion works. Eq. 4-7 shows the results corresponding to the routes illustrated in the figure.

　膨張仕事は，気体の最終膨張状態だけでなく膨張過程にも依存する．ここで，図4-3に示すような異なる膨張の道筋を検討する．膨張の間の気体の圧力は，膨張の道筋 Y_{route} ごとに P_a, P_b, P_c（式（4-6））となる．式（4-1）から膨張仕事の比較を簡単に行うことができ，図に示した道筋に対応する比較結果が式（4-7）である．

$$P_a = Y_{route-a}(V), \quad P_b = Y_{route-b}(V), \quad P_c = Y_{route-c}(V) \tag{4-6}$$

$$W_{route-a}^G < W_{route-b}^G < W_{route-c}^G \tag{4-7}$$

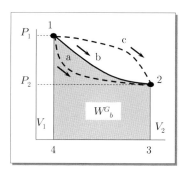

Fig.4-3　Expansion work depending on an expansion process
膨張過程に依存する膨張仕事

4.2 Internal Energy

4.2 内部エネルギー

It is clear that expansion work depends on the route of the expansion process. It means that

　膨張仕事が膨張の道筋に依存していることは明らかな事実である．このことは，図4-3

the expansion work illustrated in Fig.4-3 is not defined by the initial and final states of expansion process. In other words, work is not a kind of thermodynamic state quantity such as temperature and pressure.

In order to clarify the heat balance in the expansion process of a gas, the concept of internal energy should be introduced using the definition of Eq. 4-8.

において，膨張仕事は膨張過程の最初と最後の状態から規定されないことを意味している．言い換えれば，仕事は温度や圧力のように物体の熱力学的状態を定めている状態量ではないことを示している．

気体の膨張過程におけるエネルギーバランスを明らかにするためには，式（4-8）で定義される内部エネルギーの概念を導入する必要がある．

$$U_i = G \int_0^i c_v dT + U_0 \qquad (4\text{-}8)$$

As shown in Fig 4-4, expansion work is generally accompanied with heat movement which should be considered in the heat balance during the process. Using the definition of internal energy and considering the empirical fact of expansion work, energy balance is obtained as indicated in Eq. 4-9.

図 4-4 に示すように，膨張仕事は一般的には熱の移動をともなっている．そこで膨張過程の間の熱の移動を熱バランスの際に考慮する必要がある．その結果，内部エネルギーの定義および膨張仕事に関する経験的な事実から，式（4-9）に示されるようなエネルギーバランスの式が得られている．

$$Q_{1\text{-}2} = U_2 - U_1 + W^G \qquad (4\text{-}9)$$

This equation gives the relationship between heat, internal energy, and work. Considering a small increment in the process, the energy balance of a gas of unit mass and its total differential expression can be obtained as follows:

この式は，熱，内部エネルギー，仕事の三者の間の関係を与える式となる．微小な膨張過程の一部分の微少変化を考えれば，単位質量の気体についてのエネルギーバランスやその全微分は，以下のように求めることができる．

$$\Delta Q = \Delta U + \Delta W^G \qquad (4\text{-}10)$$
$$\Delta q = \Delta u + \Delta W \qquad (4\text{-}11)$$
$$dq = du + dW \qquad (4\text{-}12)$$

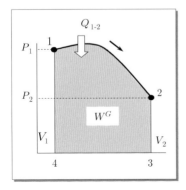

Fig.4-4 Expansion work and heat
膨張仕事と熱

4.3 Technical Work

Using the illustration of Fig.4-5, we consider the working process of a gas flowing into a cylinder, expanding within, and flowing out. All the processes are noted with 1-2-3-4 on the P-V chart in the figure.

In the process of charging (4-1), work is done by the gas to the piston as noted in Eq. 4-13.

$$W_{4-1}^{G} = P_1 A_1 x_1 = P_1 V_1 \qquad (4\text{-}13)$$

After that, the gas expands from state 1 to state 2, and the work done by the gas on the piston is described in Eq. 4-14

$$W_{1-2}^{G} = \int_1^2 P dV \qquad (4\text{-}14)$$

During the discharge process expressed by 2-3, the piston does work on the gas to expel it out of the cylinder. It indicates a negative work done by the gas and can be expressed as follows:

4.3 工業仕事

図 4-5 に示した説明図を用いて，シリンダに流入してシリンダ内で膨張し，その後流出するガスの仕事過程を検討する．このすべての過程は，説明図の P-V 線図上で，1-2-3-4 の過程として記載されている．

流入過程（4-1）では，気体がピストンにする仕事は式（4-13）により記述される．

その後，気体は状態 1 から状態 2 へ膨張するので，気体がピストンにする仕事は式（4-14）により記述される．

図の 2-3 で示されている気体の流出過程では，気体を流出させるためにピストンが気体に仕事をすることになる．このことは，気体による負の仕事を意味していて，つぎの式のように表現される．

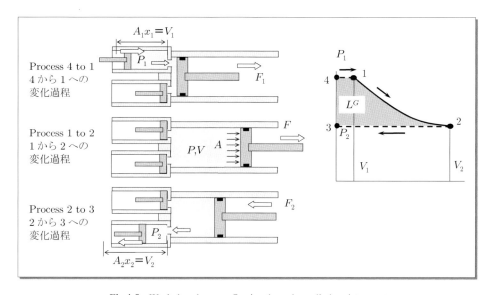

Fig.4-5 Work done by a gas flowing through a cylinder-piston system
シリンダ-ピストン系を通過する気体による仕事

$$W_{2\text{-}3}^{G} = -P_2 A_2 x_2 = -P_2 V_2 \tag{4-15}$$

As for the work done by the gas flowing through the cylinder-piston system, all of the three works mentioned above should be taken into account.

This work has a different meaning from that of simple expansion work. It is called technical work L^{G}, because it is important in the thermo-dynamic treatment for energy conversion. For the case of Fig.4-5, it can be derived as follows:

シリンダ-ピストン系を通過するこの気体が行う仕事としては，上記の三つの仕事すべてを考慮しなければならない．

この仕事は単純な膨張仕事と異なる内容を含んでいる．さらに，この仕事はエネルギー変換の場合の熱力学的取り扱いにとって非常に重要であり，そのため，工業仕事 L^{G} とよばれている．図 4-5 の場合では，これはつぎのように求めることができる．

$$L_{4\text{-}1\text{-}2\text{-}3}^{G} = W_{4\text{-}1}^{G} + W_{1\text{-}2}^{G} + W_{2\text{-}3}^{G} = P_1 V_1 + \int_{1}^{2} P dV - P_2 V_2 = \text{Area}(4\text{-}1\text{-}2\text{-}3) \tag{4-16}$$

It is worth to note that technical work is ex-pressed by the area (4-1-2-3). Using a specific volume, the technical work L done by a gas of unit mass is expressed as follows:

ここで，工業仕事は面積（4-1-2-3）で表されていることに留意すべきである．比容積を用いれば，単位質量の気体によって行われる工業仕事は以下のように表すことができる．

$$L^{G} = -\int_{1}^{2} V dP \tag{4-17}$$

$$L^{G} = -G\int_{1}^{2} v dP, \quad L = -\int_{1}^{2} v dP \tag{4-18}$$

Here, the negative sign of the integral means that the integration direction is the inverse direc-tion of a normal increment of pressure.

ここで，積分の負の符号は，圧力の通常の増加方向と逆の方向に積分が行われていることを意味している．

4.4 Enthalpy

A process that produces technical work is also coupled with heat movement as shown in Fig.4-6. Here, in order to consider an energy

4.4 エンタルピー

工業仕事を生じる熱力学的過程もまた，図4-6 に示すように熱の移動をともなっている．ここで，工業仕事に関するエネルギーバラン

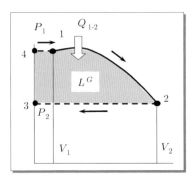

Fig.4-6   Technical work and heat
工業仕事と熱

balance concerning the technical work, a concept of enthalpy defined by Eq. 4-19 is introduced.

スを検討するため，式（4-19）で定義される
エンタルピーの概念を導入する．

$$H_i = G\int_0^i c_p dT + H_0 \qquad (4\text{-}19)$$

Using the assumption of ideal gas and Eq. 4-20, the enthalpy can be noted as follows:

理想気体を想定し，式（4-20）を使うことにより，エンタルピーは以下のように記述できる．

$$c_p = c_v + R, \quad PV = GRT \qquad (4\text{-}20)$$

$$H_i = G\int_0^i (c_v + R)dT + H_0 \qquad (4\text{-}21)$$

$$H_i = G\int_0^i c_v dT + G\int_0^i R dT + H_0 \qquad (4\text{-}22)$$

According to this definition of enthalpy, the energy balance during the change from state 1 to state 2 can be described by Eq. 4-24.

このエンタルピーの定義によれば，状態1から状態2への熱力学的変化の間のエネルギーバランスは，式（4-24）で記述することができる．

$$\begin{aligned}H_2 - H_1 &= U_2 - U_1 + \int_1^2 d(PV)\\ &= U_2 - U_1 + \int_1^2 PdV + \int_1^2 VdP\\ &= U_2 - U_1 + W^G - L^G\\ &= Q_{1\text{-}2} - L^G\end{aligned} \qquad (4\text{-}23)$$

$$Q_{1\text{-}2} = H_2 - H_1 + L^G \qquad (4\text{-}24)$$

Finally, the following equations can be derived from the definition of enthalpy.

したがって，最終的にエンタルピーの定義より以下の式を導出することができる．

$$\Delta Q = \Delta H + \Delta L^G \qquad (4\text{-}25)$$

$$\Delta q = \Delta h + \Delta L \qquad (4\text{-}26)$$

$$dq = dh + dL \qquad (4\text{-}27)$$

4.5 The First Law of Thermodynamics

4.5 熱力学の第一法則

The first law of thermodynamics is the energy balance among internal energy, heat, and work. As shown in Fig.4-7, a thermodynamic process from state 1 to state 2 is always accompanied with heat movement or work. Thermodynamic quantities such as pressure, specific volume, temperature, internal energy, and enthalpy

熱力学の第一法則は，内部エネルギーと熱と仕事の三者の間に成立するエネルギーバランスである．図4-7に示したように，状態1から状態2への熱力学的変化過程では，熱の移動または仕事がつねに生じている．圧力，比容積，温度，内部エネルギー，エンタルピーのような熱力学的状態量は，その過程ととも

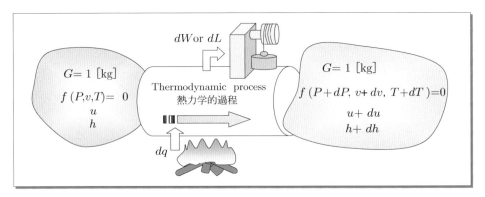

Fig.4-7   The first law of thermodynamics   熱力学の第一法則

changes simultaneously with the process. By considering the heat and work with regard to the process, we can clearly derive the relationships among the thermodynamic quantities of state. The main equations are as follows:

に変化する．このプロセスに関する熱と仕事を考慮することにより，熱力学的状態量の間の関係を明確に明らかにすることが可能である．関係式の主要なものを以下に記述しておく．

$$dq = du + dW = dh + dL \tag{4-28}$$

$$dq = du + Pdv \tag{4-29}$$

$$q_{12} = u_2 - u_1 + \int_1^2 Pdv = u_2 - u_1 + W_{12} \tag{4-30}$$

$$dh = du + d(Pv) = du + Pdv + vdP = dq + vdP \tag{4-31}$$

$$dq = dh - vdP \tag{4-32}$$

$$q_{12} = h_2 - h_1 - \int_1^2 vdP = h_2 - h_1 + L_{12} \tag{4-33}$$

Problems

△4-1

Determine each expansion work when a gas expands from state 1 to state 2 along process paths A, B, and C, as shown in Fig. 4-8.

△4-2

Determine each technical work when a gas expands from state 1 to state 2 along process paths A, B, and C, as shown in Fig. 4-8.

△4-3

A frictionless piston and cylinder device (see Fig.4-9) initially contains air of 1 m³ at 0.25 MPa. Now 2 MJ of heat is transferred to the air and the

問題

△4-1

図 4-8 に示すように，気体が経路 A, B, C に沿って状態 1 から状態 2 に変化するとき，それぞれの膨張仕事を求めなさい．

△4-2

図 4-8 に示すように，気体が経路 A, B, C に沿って状態 1 から状態 2 に変化するとき，それぞれの工業仕事を求めなさい．

△4-3

摩擦のないピストンとシリンダからなる装置（図 4-9）の内部に，初期状態で 0.25 MPa の空気が 1 m³ 入っている．その空気に 2 MJ の熱を加え

volume of air increases to 4 m³. During this process the pressure remains constant. Determine the change in internal energy of air in this process.

たところ，その体積が 4 m³ に増加した．この過程で圧力は一定のままであった．このとき空気の内部エネルギーの変化を求めなさい．

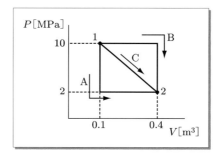

Fig.4-8 *P-V* Chart for Problem 4-1　問題 4-1 の *P-V* 線図

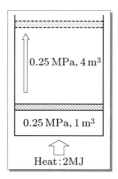

Fig.4-9 Schematic for Problem 4-3　問題 4-3 の説明図

Chapter 5
Internal Energy and Enthalpy
内部エネルギーとエンタルピー

5.1 Internal Energy and Quantity of State

5.1 内部エネルギーと状態量

Using the equation of state for ideal gas and the fundamental equation for internal energy, we consider the thermodynamic process shown in Fig.5-1.

理想気体の状態方程式や内部エネルギーについての基本式を基に，図 5-1 に示した熱力学的過程を検討する．

$$PV = GRT \tag{5-1}$$

$$dQ = dU + dW^G \tag{5-2}$$

$$U_2 = U_1 + \int_1^2 dQ - \int_1^2 dW^G \tag{5-3}$$

According to the fact that the internal energy is a kind of thermodynamic quantity of state, the internal energy of state 2 should be defined independently from the routes or processes. Then, Eq. 5-4 can be derived.

内部エネルギーが状態量の一つであるということから，状態 2 の内部エネルギーは変化の道筋とは無関係に規定される．その結果として式（5-4）が得られる．

$$U_2 = U_{2(a)} = U_{2(b)} = U_{2(c)} = G\int_1^2 c_v dT + U_0 \tag{5-4}$$

Furthermore, work and heat for each route should satisfy the following equations:

さらに，それぞれの道筋に対応する熱と仕事については，つぎの式が成立する．

$$\int_{1-a-}^2 dQ - \int_{1-a-}^2 dW^G = \int_{1-b-}^2 dQ - \int_{1-b-}^2 dW^G = \int_{1-c-}^2 dQ - \int_{1-c-}^2 dW^G \tag{5-5}$$

$$\int_{1-a-}^2 dQ + \int_{2-b-}^1 dQ = \int_{1-a-}^2 dW^G + \int_{2-b-}^1 dW^G \tag{5-6}$$

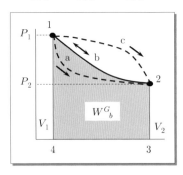

Fig.5-1 Thermodynamic process from state 1 to state 2
状態 1 から状態 2 への熱力学的変化

Here,by considering a cyclic process noted by 1-a-2-b-1, we can derive Eq. 5-7 and another form of the first law of thermodynamics, i. e., heat and work are equivalent and interchangeble.

ここで，図 5-1 に示すように 1-a-2-b-1 で記述される一回りの過程を考えることにより，式（5-7）が得られ，熱力学の別の表現形式を得ることができる．すなわち，熱と仕事は同等であり，両者は互いに変換可能であるということを導きだすことができる．

$$\oint_{1-a-2-b-1} dQ = \oint_{1-a-2-b-1} dW^G \qquad (5\text{-}7)$$

This integral equation means that the internal energy at state 1 is uniquely fixed as a state quantity, because both the summation of heat and that of work in a cycle process are canceled with each other.

サイクルにおける熱の合計と仕事の合計が打ち消し合うことから，この積分は状態 1 の内部エネルギーが状態量として一つの値に定まることを意味している．

5.2 Pressure and Kinetic Energy of Molecules

5.2 圧力と分子の運動エネルギー

Consider gaseous molecules moving in a cubic container of size "a" in an x-y-z coordinate as shown in Fig.5-2. The velocity of i-th molecule is expressed in Eq. 5-8, using three velocity components.

図 5-2 に示すような，寸法 a の立方体容器中で運動している粒子と x-y-z の座標系を想定する．i 番目の分子の速度は，座標系に対応した三つの速度成分を用いて式（5-8）のように記述することができる．

$$w_i^2 = w_{xi}^2 + w_{yi}^2 + w_{zi}^2 \qquad (5\text{-}8)$$

Assuming an elastic collision between the molecule and the wall, the momentum change at every collision can be noted as $2\,m_i v_i$. Thus, the force/second given by a molecule is obtained by Eq. 5-9

分子と壁との間に弾性衝突の過程が成り立つとすれば，毎回の衝突による運動量の変化は $2\,m_i v_i$ となる．したがって，一つの分子が与える力/秒は，式（5-9）のようになる．

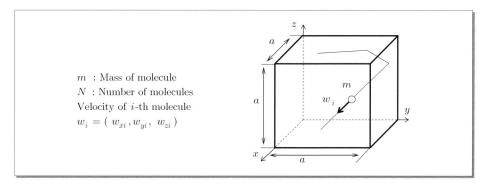

m : Mass of molecule
N : Number of molecules
Velocity of i-th molecule
$w_i = (\,w_{xi}, w_{yi},\ w_{zi}\,)$

Fig.5-2　Molecular dynamics and pressure　分子の運動力学と圧力

$$\dot{F}_{xi} = \frac{w_{xi}}{2a} \cdot 2mw_{xi} = \frac{mw_{xi}^2}{a} \tag{5-9}$$

Considering the number of molecules and the wall area, the pressure caused by molecular collisions can be noted as follows:

分子の総数と壁の面積を考慮すると，分子衝突に起因する圧力は，以下のように記述することができる．

$$P_x = \frac{1}{a^2 [\mathrm{m}^2]} \sum_i^N \dot{F}_{xi} [\mathrm{N/s}] \cdot 1 [\mathrm{s}] = \frac{1}{a^2} \sum_{i=1}^N \frac{mw_{xi}^2}{a} = \frac{1}{a^3} \sum_{i=1}^N mw_{xi}^2 \tag{5-10}$$

$$P_y = \frac{1}{a^3} \sum_{i=1}^N mw_{yi}^2, \qquad P_z = \frac{1}{a^3} \sum_{i=1}^N mw_{zi}^2 \tag{5-11}$$

From the law of energy equipartition, the following three equations can be derived:

エネルギーの等分配法則からは，以下の三つの式が得られる．

$$\sum_{i=1}^N w_i^2 = \sum_{i=1}^N w_{xi}^2 + \sum_{i=1}^N w_{yi}^2 + \sum_{i=1}^N w_{zi}^2 \tag{5-12}$$

$$\sum_{i=1}^N w_{xi}^2 = \sum_{i=1}^N w_{yi}^2 = \sum_{i=1}^N w_{zi}^2 \tag{5-13}$$

$$\overline{w}^2 = \frac{1}{N} \sum_{i=1}^N w_i^2 = \frac{3}{N} \sum_{i=1}^N w_{xi}^2 = \frac{3}{N} \sum_{i=1}^N w_{yi}^2 = \frac{3}{N} \sum_{i=1}^N w_{zi}^2 \tag{5-14}$$

Where, $\overline{w}$ is the average velocity of molecules. Then, the pressure caused by molecular collisions is expressed as follows:

ここで，$\overline{w}$ は分子の平均速度である．以上より，分子の衝突に起因する圧力は以下のようになる．

$$P = P_x = P_y = P_z = \frac{1}{a^3} \sum_{i=1}^N mw_{xi}^2 = \frac{1}{a^3} \cdot \frac{mN\overline{w}^2}{3} \tag{5-15}$$

Using the definition of molecular kinetic energy and the container volume, we can finally obtain the following equations:

分子の運動エネルギーの定義や容器の容積により，最終的に以下のような式を得ることができる．

$$e = \frac{1}{2} m\overline{w}^2 \quad (\text{kinetic energy of molecule 分子の運動エネルギー}) \tag{5-16}$$

$$V = a^3 \quad (\text{volume of container 容器の容積}) \tag{5-17}$$

$$PV = mN \cdot \frac{\overline{w}^2}{3} \tag{5-18}$$

$$PV = \frac{2}{3} N \cdot e = GRT \tag{5-19}$$

Using the number of molecules in the container, the internal energy of gas can be noted as

容器内の分子の数を考慮すれば，気体の内部エネルギーは，つぎのように記述される．

follows:

$$U = N \cdot e = \frac{3}{2} GRT \qquad (5\text{-}20)$$

Furthermore, from a thermodynamic definition, the internal energy can be also noted by Eq. 5-21,

さらに，熱力学における定義から，内部エネルギーは式（5-21）のように記述することができる．

$$U = G \int_0^T c_v dT = G c_v T \qquad (5\text{-}21)$$

From a comparison of Eq. 5-20 and 5-21, we can obtain the important relationships among specific heat, gas constant, and the ratio of specific heats.

式（5-20）と式（5-21）の比較から，比熱，ガス定数，比熱比の三者の間に成立する重要な，つぎの関係式を得ることができる．

$$c_v = \frac{3}{2} R, \quad \kappa = \frac{c_p}{c_v} = \frac{c_v + R}{c_v} = \frac{5}{3} \quad \text{(for a single atom molecule　単原子分子)} \qquad (5\text{-}22)$$

In the analysis presented above, only the molecular energy of translation is considered. However, the total molecular energy contains both rotation and vibration energies of molecules as shown in the illustrations of Fig.5-3 and Eq. 5-23.

上記の解析では，並進運動としての分子のエネルギーのみを考慮している．しかし，図5-3 のイラストレーションや式（5-23）のように，分子のエネルギーには，分子の回転や振動が含まれている．

$$e_{molecule} = e_{translation} + e_{rotation} + e_{vibration} \qquad (5\text{-}23)$$

As for the case of a single atom molecule, no rotation and no vibration mode of freedom is needed. Internal energy of vibration mode is usually not large compared to the translation energy and is negligible for thermodynamic analysis. As a result, 5 degrees of freedom for a two atom molecule and 6 degrees of freedom for a three atom molecule should be considered in internal energy calculations. In other words, it should be considered that twice amount of ki-

単原子分子の場合では，分子の回転や振動の自由度を考慮する必要はない．振動モードの内部エネルギーは，並進運動のエネルギーに比べて大きくないので，熱力学的な解析では省略することが可能である．結果として，二原子分子では五つの自由度，三原子分子では六つの自由度を内部エネルギーの算出において考慮する必要が生じる．いいかえると，このことは単原子分子（$n^* = 3$）にくらべて三原子分子（$n^* = 6$）は 2 倍の運動エネルギー

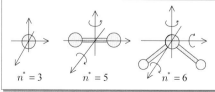

Degree of freedom (translation + rotation)
$n^* = 3$　for single atom molecule (3 translation)
$n^* = 5$　for two atom molecule (3 translation + 2 rotation)
$n^* = 6$　for three atom molecule (3 transration + 3 rotaion)

$n^* = 3$　　$n^* = 5$　　$n^* = 6$

Fig.5-3　Degree of freedom of a molecule　分子の自由度

netic energy can be belonged by three atom molecule ($n^* = 6$)compared to single atom molecule ($n^* = 3$). Since the law of energy equipartition is applicable to both translation and rotation energies, Eq. 5-24 and 5-25 can be obtained. The values of κ are already shown by Eq. 3-11.

を保持できることを考慮する必要があるということである．エネルギーの等分配法則が並進運動と回転の両者に適応されるので，結果として式（5-24）と式（5-25）が得られる．この κ の値は式(3-11)にてすでに示してある．

$$c_v = \frac{n^*}{2}R = \frac{5}{2}R, \quad \kappa = \frac{c_p}{c_v} = \frac{c_v + R}{c_v} = \frac{7}{5} \quad \text{(for a two atom molecule 二原子分子)} \quad (5\text{-}24)$$

$$c_v = \frac{n^*}{2}R = \frac{6}{2}R, \quad \kappa = \frac{c_p}{c_v} = \frac{c_v + R}{c_v} = \frac{4}{3} \quad \begin{array}{l}\text{(for a poly ($\geq$3) atomic molecule} \\ \text{多原子分子)}\end{array} \quad (5\text{-}25)$$

5.3 Charge Work and Enthalpy

5.3 充填仕事とエンタルピー

Consider a gas charging process shown in Fig.5-4. G_1 is the initial mass of gas in the cylinder and G_p is the mass of additional gas to be charged. As shown in the figure, the pressure of the system does not change through all the processes and also no temperature difference exists.

The work of gas charged into the cylinder is noted by Eq. 5-26.

図 5-4 に示す気体を充填する過程を検討する．G_1 はシリンダ内に最初からある気体の質量であり，G_p は充填される補充気体の質量である．図に示すように，この系の圧力は充填の全過程を通じて変化せず，また，温度差も生じていないとする．

シリンダ内に充填される気体の仕事は，式（5-26）で記述される．

$$\Delta W_p^G = P \cdot A_p x_p = P \cdot V_p \tag{5-26}$$

From the definition of enthalpy, the following equations are applicable for each gas.

エンタルピーの定義から，それぞれの気体について，以下のような式が適用される．

$$H_1 = U_1 + PV_1, \quad H_p = U_p + PV_p \tag{5-27}$$

$$H_2 = U_2 + PV_2 = U_2 + P(V_1 + V_p) \tag{5-28}$$

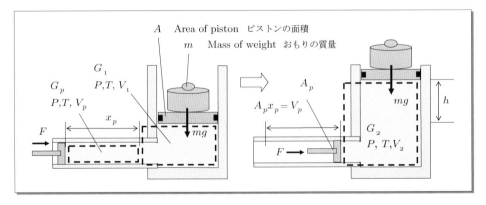

Fig.5-4 Enthalpy change during gas charging 気体の充填時のエンタルピー変化

$$H_2 = U_2 + PV_2 = U_1 + U_p + PV_2 = (H_1 - PV_1) + (H_p - PV_p) + PV_2 = H_1 + H_p \qquad (5\text{-}29)$$

From the energy balance of the system, energy E_2 of final state can be expressed using the internal energy of gas and the increase in the weight's potential energy.

系のエネルギーバランスから，最終状態のエネルギー E_2 は，気体の内部エネルギーとおもりの位置エネルギーの増加により表すことができる．

$$E_2 = E_1 + E_p + PV_p = E_1 + E_p + \Delta W_p^G = U_2 + mgh \qquad (5\text{-}30)$$

$$E_2 - E_1 = E_p + PV_p = U_2 - U_1 + PV_p = U_p + PV_p = H_p \qquad (5\text{-}31)$$

Thus, the energy increase of the system during gas charging becomes equivalent to the enthalpy of the additional gas to be charged. Further, the energy increase is larger than the internal energy increase of charged gas, and it is caused by the piston work.

したがって，気体の充填の前後のエネルギー増加は，充填するための補充気体のエンタルピーと等しくなる．さらに，このエネルギーの増加は，充填されたガスの内部エネルギーの増加より大きくなる．また，それはピストンによってなされた仕事のためである．

$$\Delta E_{12} = H_p = U_p + \Delta W_p^G \qquad (5\text{-}32)$$

5.4 Enthalpy and Flow

5.4 エンタルピーと流れ

A flow system that has inlet and outlet ports for flowing media has become one of the engineering topics of thermodynamics. As shown in Fig.5-5, the flowing media is passing through the system and is not stored inside the system. Then, the flow rate expressed by Eq. 5-33 should be used instead of the extensive quantity of ther-

流体物質に対する流入口と流出口をもつ流れの系は，熱力学に関連する工学技術上の一つの検討事項になっている．図5-5に示すとおり，流動する物質は対象とする系を通過し，系内に留まることはない．そこで，熱力学における示量性量の代わりに，式（5-33）で示されている流量を用いる必要が生じてくる．

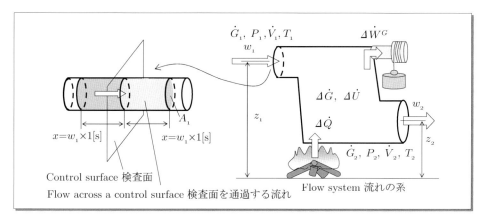

Fig.5-5 Flow system and control surface of flow　流れの系と流れの検査面

modynamics.

$$\dot{Y} = \frac{dY}{dt} \left[\dot{Y}: \text{Rate of } Y, \text{ ex.: } \dot{G} \text{ means mass flow rate 質量流量 [kg/s]} \right] \qquad (5\text{-}33)$$

For example, when the volumetric flow rate and the mass flow rate at the inlet control surface are known, an equation of state using the volumetric and mass flow rates can be used instead of the ordinary equation of state.

たとえば，入口の検査面において体積流量と質量流量が既知となっているのであれば，通常の状態方程式に代えて，体積流量と質量流量を用いた状態方程式を使うことが可能になる．

$$PV = GRT \Rightarrow P\dot{V} = \dot{G}RT \qquad (5\text{-}34)$$

As for the case of Fig.5-5, since the inlet velocity of flow media is known, the mass flow rate and the enthalpy flow rate can be noted as follows:.

図 5-5 の例では，流体物質の流入速度が既知であるため，質量流量とエンタルピーの流入量は以下のように記述することができる．

$$\dot{G}_1 = \frac{dG_1}{dt} \times 1 = \frac{P_1 \dot{V}_1}{RT_1} = \frac{P_1 \times A_1 w_1}{RT_1} \qquad (5\text{-}35)$$

$$\frac{dH_1}{dt} \times 1 = \frac{dG_1}{dt} \times 1 \times (u_1 + P_1 v_1) = \dot{G}_1 (u_1 + P_1 v_1) = \dot{U}_1 + P_1 \dot{V}_1 = \dot{H}_1$$
$$\text{(Enthalpy/(unit time) flowing into a system}$$
$$\text{系に流入する単位時間あたりのエンタルピー)} \qquad (5\text{-}36)$$

The energy supplied through the inlet control surface can be noted using the flow rate of internal energy or the flow rate of enthalpy.

入口の検査面を通過して供給されるエネルギーは，内部エネルギーまたはエンタルピーの流入量を用いて記述することができる．

$$\dot{E}_1 = \dot{U}_1 + P_1 \dot{V}_1 + \frac{1}{2} \dot{G}_1 w_1^2 + \dot{G}_1 g z_1 \qquad (5\text{-}37)$$

$$\dot{E}_1 = \dot{H}_1 + \frac{1}{2} \dot{G}_1 w_1^2 + \dot{G}_1 g z_1 \qquad (5\text{-}38)$$

As for the energy released through the outlet control surface, the following equations are obtained:

出口における検査面から流出するエネルギーについては，つぎのような式が成り立つ．

$$\dot{E}_2 = \dot{U}_2 + P_2 \dot{V}_2 + \frac{1}{2} \dot{G}_2 w_2^2 + \dot{G}_2 g z_2 \qquad (5\text{-}39)$$

$$\dot{E}_2 = \dot{H}_2 + \frac{1}{2} \dot{G}_2 w_2^2 + \dot{G}_2 g z_2 \qquad (5\text{-}40)$$

When additional sources of mass and heat are included in the system, it should be treated as external sources of heat and work, respectively. Then, the mass and energy balance of the system are summed up as follows:

もし，質量や熱の発生源が系内にあるならば，それは外部からの加熱や外部への仕事とともに取り扱う必要がある．したがって，系の質量とエネルギーのバランスはつぎの各式のようにまとめることができる．

$$\dot{G}_1 + \Delta \dot{G} = \dot{G}_2 \tag{5-41}$$

$$\dot{E}_1 + \Delta \dot{Q} + \Delta \dot{U} = \dot{E}_2 + \Delta \dot{W}^G \tag{5-42}$$

$$\dot{U}_1 + P_1 \dot{V}_1 + \frac{1}{2} \dot{G}_1 w_1^2 + \dot{G}_1 g z_1 + \Delta \dot{Q} + \Delta \dot{U} = \dot{U}_2 + P_2 \dot{V}_2 + \frac{1}{2} \dot{G}_2 w_2^2 + \dot{G}_2 g z_2 + \Delta \dot{W}^G \tag{5-43}$$

As shown in Fig.2-6, both kinetic and potential energies of the fluid are so small compared to the internal energy or enthalpy. Thus, Eq. 5-43 can be reduced to Eq. 5-44 and this relationship is known as the iso-enthalpy flow of fluid.

図 2-6 で示したように，流体の運動エネルギーや位置エネルギーは，内部エネルギーやエンタルピーに比べてたいへん小さな量である．したがって，式（5-43）は式（5-44）のように簡略化することができ，この関係は流体の等エンタルピー流れとして知られている．

$$\frac{\frac{1}{2}\dot{G}w^2 + \dot{G}gz}{\dot{U}} \approx 0 \Rightarrow \begin{cases} \dot{U}_1 + P_1 \dot{V}_1 + \Delta \dot{Q} + \Delta \dot{U} = \dot{U}_2 + P_2 \dot{V}_2 + \Delta \dot{W}^G \\ \dot{H}_1 + \Delta \dot{Q} + \Delta \dot{U} = \dot{H}_2 + \Delta \dot{W}^G \end{cases} \tag{5-44}$$

5.5 Bernoulli's Equation

Consider a special case in which no internal and external heat sources and no internal increase of mass exist, and furthermore, no external work is required. With these assumptions, the flow system shown by Fig.5-5 can be simplified as indicated in Fig.5-6.

Using the conditions mentioned above and considering an isothermal system, Eq. 5-43 can be reduced as indicated by Eq. 5-47.

5.5 ベルヌーイの式

内部および外部の熱源がなく，内部での質量増加が起きず，さらに外部仕事も必要とされていない特別な場合を検討する．このような仮定によれば，図 5-5 に示した流れの系は，図 5-6 のように簡略化される．

上記の条件に等温系であるという条件を付け加えると，式（5-43）は式（5-47）に示すように簡略化される．

$$\Delta \dot{Q} = 0, \quad \Delta \dot{U} = 0, \quad \Delta \dot{G} = 0, \quad \Delta \dot{W}^G = 0 \tag{5-45}$$

$$T_1 = T_2 \tag{5-46}$$

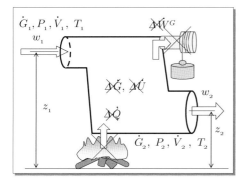

Fig.5-6 Isothermal flow for bernoulli's equation
　　　　　ベルヌーイの式に対応する等温流れ

$$P_1\dot{V}_1 + \frac{1}{2}\dot{G}_1 w_1^2 + \dot{G}_1 g z_1 = P_2\dot{V}_2 + \frac{1}{2}\dot{G}_2 w_2^2 + \dot{G}_2 g z_2 \tag{5-47}$$

Similar to fluid dynamics, the density of fluid defined by Eq. 5-48 can be used instead of the specific volume of the fluid. Then, Bernoulli's equation shown by Eq. 5-49 can be obtained.

流体力学の場合と同様，式（5-48）で定義されている流体の密度を流体の比容積の替わりに使うことができるので，式（5-49）に示したベルヌーイの式を導くことができる．

$$\rho_1 = \frac{\dot{G}_1}{\dot{V}_1}, \quad \rho_2 = \frac{\dot{G}_2}{\dot{V}_2} \tag{5-48}$$

$$P_1 + \frac{1}{2}\rho_1 w_1^2 + \rho_1 g z_1 = P_2 + \frac{1}{2}\rho_2 w_2^2 + \rho_2 g z_2 \tag{5-49}$$

Problems   問題

△5-1

Derive Eq. 5-50 and determine the mean velocities $\sqrt{\overline{w^2}}$ of H_2 and CO_2 at 300 K.

△5-1

式（5-50）を導出し，300 K における H_2 と CO_2 の平均速度 $\sqrt{\overline{w^2}}$ を求めなさい．

$$\sqrt{\overline{w^2}} = \sqrt{3RT} \tag{5-50}$$

△5-2

Air is compressed steadily with a compressor. The mass flow rate of the air is 0.3 kg/s. The specific enthalpy and air velocity at the inlet are 250 kJ/kg and 40 m/s respectively, and those at the outlet are 400 kJ/kg and 100 m/s respectively. If a heat loss of 7 kJ/kg per second occurs during the process, determine the necessary power input to the compressor. Here, no potential energy change is assumed.

△5-2

空気が圧縮機で定常的に圧縮されている．質量流量は 0.3 kg/s である．圧縮機入口における比エンタルピーと流速はそれぞれ 250 kJ/kg，40 m/s で，出口における比エンタルピーと流速はそれぞれ 400 kJ/kg，100 m/s である．圧縮の過程で熱損失が毎秒 7 kJ/kg あるとするとき，圧縮機の運転に必要な動力を求めなさい．ここで，位置エネルギーの変化はないと仮定する．

△5-3

A liquid in a tank is ejected through a small hole

△5-3

タンクに入っている液体が，図 5-7 のようにタ

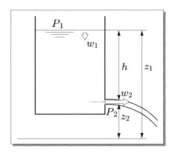

Fig.5-7   Ejected liquid through a small hole for Problem 5-3
　　　　　　　問題 5-3 にて小噴孔から放出される液体

as shown Fig.5-7. Here the area of hole is much smaller than the area of cross section of the tank, and therefore $w_1 \ll w_2$. In this case, determine the liquid ejection velocity w_2.

ンクに開けられた小さな穴から噴出している．ここで，穴の面積はタンクの断面積より非常に小さく，$w_1 \ll w_2$ が成り立っている．このとき，穴からの噴出速度 w_2 を求めなさい．

Chapter 6
The Second Law of Thermodynamics
熱力学の第二法則

6.1 Concept of Entropy

6.1 エントロピーの概念

Consider an isolated thermodynamics system that is composed of high and low temperature heat substances. The high temperature substance has enough heat capacity and it is assumed that no temperature decrease occurs during heat discharge. This means that the high temperature heat source has an infinitely large heat capacity. The low temperature substance acts as a low temperature heat sink. The quantity of heat discharged from the high temperature source is received by a low temperature heat source (sink) after passing through a thermodynamic process.

From the first law of thermodynamics, in other words, the law of energy conservation, the

　高温と低温の物体から構成されている孤立した熱力学的な系を検討する. 高温の物体は, 十分な熱容量をもち, 熱の放出の際に温度低下は起こらないものと想定する. このことは, 無限大の熱容量の高温熱源を意味している. 低温の物体は, 低温の熱の吸収源として作用する. 高温熱源から放出された熱量は, 熱力学的な過程を経た後, 低温熱源 (低温吸熱源) に受け入れられる.

　熱力学の第一法則, 言い換えればエネルギーの保存則より, 低温の熱の吸収源が受け

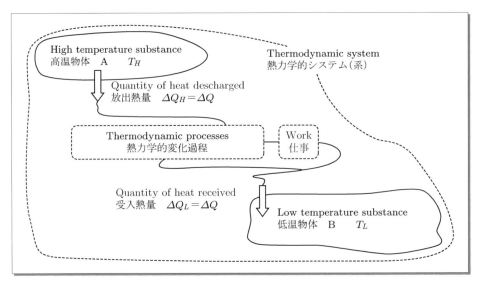

Fig.6-1　Thermodynamic system and transport phenomena of heat
　　　　　熱力学的システムと熱の移動現象

quantity of heat received by the low temperature heat sink is the same as the quantity of discharged heat as shown in Eq. 6-1. Even though a part of heat is converted to a work, it is finally re-converted to heat by other irreversible process such as friction. Then, the energy conservation shown in Fig.6-1 is generally maintained.

取った熱量は，式 (6-1) のように放出された熱量と等しい．たとえば，熱の一部が仕事に変換されたとしても，それは摩擦のような非可逆過程で最終的に熱にもどされる．したがって，図6-1に示したエネルギー保存はつねに成り立っている．

$$\Delta Q_H = \Delta Q_L = \Delta Q \quad \text{(conservation of energy　エネルギーの保存)} \tag{6-1}$$

The thermodynamic system considered here is isolated, and no external work is performed from outside of the system. Thus, Eq. 6-2 is given as an energy conservation equation of the system.

ここで検討している熱力学的システムは孤立系であるので，外部からはなんらの仕事もなされていない．したがって，系のエネルギー保存式として，式 (6-2) が与えられる．

$$\Delta Q_{system} = (-|\Delta Q_H|) + (+|\Delta Q_L|) = \Delta Q_L - \Delta Q_H = 0 \tag{6-2}$$

Here, we introduce the concept of entropy defined by Eq. 6-3 to clarify the effect of heat transport phenomena in the system.

ここで，システムの熱の移動現象の効果を明確にするために，式 (6-3) で定義されるエントロピーの概念を導入する．

$$\Delta S = \frac{\Delta Q}{T} \quad \text{(definition of entropy　エントロピーの定義)} \tag{6-3}$$

The entropy changes coupled with heat loss and heat gain can be noted as follows.

熱の放出や受領にともなうエントロピーの変化は，以下のように記述できる．

$$\Delta S_H = \frac{\Delta Q_H}{T_H} = \frac{\Delta Q}{T_H} \quad \text{(quantity of discharged entropy　放出されたエントロピーの量)} \tag{6-4}$$

$$\Delta S_L = \frac{\Delta Q_L}{T_L} = \frac{\Delta Q}{T_L} \quad \text{(quantity of recived entropy　受領されたエントロピーの量)} \tag{6-5}$$

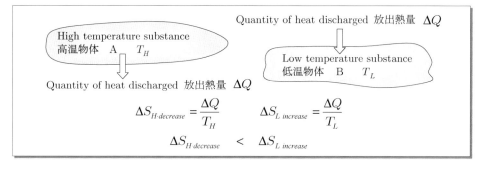

Fig.6-2　Entropy changes in high and low temperature substances in the system
熱力学的システム内の高温物体および低温物体のエントロピー変化

As a result, the overall entropy change of the system can be noted by Eq. 6-6. We can notice that the entropy of the system increased with the transport phenomena of heat.

システム全体のエントロピー変化は式 (6-6) で記述できる．したがって，熱の移動現象にともない，熱力学的な系のもつエントロピーが増加していることを明らかにすることができる．

$$\Delta S_{system} = \left(-\frac{|\Delta Q_H|}{T_H}\right)+\left(\frac{|\Delta Q_L|}{T_L}\right)= \frac{\Delta Q_L}{T_L} - \frac{\Delta Q_H}{T_H} = \Delta Q\left(\frac{1}{T_L}-\frac{1}{T_H}\right)>0 \tag{6-6}$$

Figure 6-2 shows the entropy changes of each heat source. In general, heat movement in a lower temperature range results in a larger entropy change than that of a high temperature range.

図 6-2 は熱源ごとのエントロピーの変化を示したものである．一般的に低温領域での熱の移動は高温領域における熱の移動に比べて大きなエントロピー変化をもたらすことになる．

6.2 Entropy Change

6.2 エントロピーの変化

Consider an entropy change of a gas that is contained in a cylinder-piston system as illustrated in Fig.6-3. Gas is heated at the bottom of the cylinder and expands. However, the pressure of the system is kept constant using a piston and weight mechanism. When the gas in the cylinder is assumed as an ideal gas of mass G, the equation of state and the changes of volume and temperature during the three heating processes (i = 1, 2, 3) can be noted by Eq. 6-7 and Eq. 6-8.

図 6-3 に示したシリンダ-ピストンの系内の気体のエントロピーの変化を検討する．気体はシリンダの底から加熱され膨張するが，ピストンとおもりの機構により，気体の圧力は一定に保たれている．シリンダ内の気体が質量 G の理想気体である場合では，ここでの三つの加熱過程（i = 1, 2, 3）について状態方程式や加熱過程での体積と温度の変化を式 (6-7) と式 (6-8) のように記述できる．

$$PV_i = GRT_i \quad \left(i = 1, 2, 3\right) \tag{6-7}$$

$$P = \text{const.} \Rightarrow P\,\Delta V_i = GR\,\Delta T_i \tag{6-8}$$

Using the definition of internal energy, the quantity of heat added to the gas can be noted by the change of internal energy and work to the

内部エネルギーの定義より，気体に加えられた熱は，内部エネルギーの変化とピストンへの仕事となる．

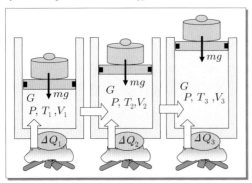

Fig.6-3 Heating process and entropy change
加熱過程とエントロピー変化

piston.

$$\Delta Q_i = \Delta U_i + P \cdot \Delta V_i = Gc_v \Delta T_i + P \cdot \Delta V_i = (Gc_v + GR) \cdot \Delta T_i = \left(\frac{Gc_v P}{GR} + P\right) \cdot \Delta V_i \qquad (6\text{-}9)$$

Then, changes of temperature and volume are expressed as follows:

この結果より，温度と体積の変化はつぎのように表現される．

$$\frac{\Delta T_i}{T_i} = \frac{1}{(Gc_v + GR)} \cdot \frac{\Delta Q_i}{T_i} = \frac{1}{Gc_p} \cdot \frac{\Delta Q_i}{T_i} \qquad (6\text{-}10)$$

$$\frac{\Delta V_i}{V_i} = \frac{GR}{Gc_v P + GR} \cdot \frac{\Delta Q_i}{V_i} = \frac{GR}{Gc_v P + GR} \cdot \frac{P}{GR} \cdot \frac{\Delta Q_i}{T_i} = \frac{P}{Gc_v P + GR} \cdot \frac{\Delta Q_i}{T_i} \qquad (6\text{-}11)$$

Here, if the temperature change during each heating process is so small as to be negligible, the entropy change of each heating process can be expressed by Eq. 6-12.

ここで，それぞれの加熱過程の間の温度変化が無視できるレベルであるとすれば，それぞれの加熱過程におけるエントロピー変化は，式（6-12）のように表すことができる．

$$\Delta S_i \quad \left[\text{Entropy エントロピー}\right] \cong \frac{\Delta Q_i}{T_i} \qquad (6\text{-}12)$$

Then, if the heat quantities for the three heating processes shown in Fig.6-3 are the same, the following relationships are obtained:

そこで，図 6-3 に示した三つの加熱過程に対する熱量が等しい場合については，以下のような式が成立する．

$$T_1 < T_2 < T_3 \quad \text{and} \quad \Delta Q_1 = \Delta Q_2 = \Delta Q_3 \qquad (6\text{-}13)$$

$$\frac{\Delta Q_1}{T_1} > \frac{\Delta Q_2}{T_2} > \frac{\Delta Q_3}{T_3} \qquad (6\text{-}14)$$

$$\frac{\Delta T_1}{T_1} > \frac{\Delta T_2}{T_2} > \frac{\Delta T_3}{T_3}, \quad \frac{\Delta V_1}{V_1} > \frac{\Delta V_2}{V_2} > \frac{\Delta V_3}{V_3} \qquad (6\text{-}15)$$

The equations above show that the effect of heat transportation in low temperature levels results in larger changes in the thermodynamic state quantities than the case of high temperature levels.

上述した式は，高温の場合に比べて低温での熱の移動が熱力学的状態量に大きな変化を与えていることを示している．

On the other hand, if the entropy changes for the three processes are equivalent as shown by Eq. 6-16, the relative changes of temperature and volume in these three processes will be equal as shown in Eq. 6-17.

一方，三つの加熱過程において，エントロピーの変化が式（6-16）のように同等であるならば，温度や体積の相対的な変化は式（6-17）に示すように三つの過熱過程において等しくなる．

$$T_1 < T_2 < T_3 \quad \text{and} \quad \frac{\Delta Q_1}{T_1} = \frac{\Delta Q_2}{T_2} = \frac{\Delta Q_3}{T_3} \qquad (6\text{-}16)$$

$$\frac{\Delta T_1}{T_1} = \frac{\Delta T_2}{T_2} = \frac{\Delta T_3}{T_3}, \quad \frac{\Delta V_1}{V_1} = \frac{\Delta V_2}{V_2} = \frac{\Delta V_3}{V_3} \qquad (6\text{-}17)$$

6.3 The Second Law of Thermodynamics

6.3 熱力学の第二法則

Most simple expressions of the second law of thermodynamics state that heat is transported from a high temperature substance to a low temperature substance. To clarify the meaning of the second law, the heat transport phenomena illustrated by Fig.6-4 is considered.

In order to release or to discharge a quantity of heat, a high temperature substance needs ambient surroundings or substances with lower temperatures. As a result of this heat release, the temperature of the substance decrease slightly. On the other hand, a low temperature substance needs a high temperature heat source to receive heat. During the heat transport process, heat itself is conserved and Eq. 6-18 always exists as a fact of energy conservation.

熱は温度の高い物体から温度の低い物体に移動するということが，熱力学の第二法則のもっとも簡潔な表現である．熱力学の第二法則の意味を明確に示すために，図6-4にあるような熱の移動現象を検討する．

ある熱量を高温の物体から解き放つか放出するためには，温度の低い周囲環境や物体が必要になる．また，熱を放出した結果として，物体の温度は幾分低下する．一方，低温の物体がある熱量を受け取るためには，高温の熱源が必要になる．熱の移動の過程において熱自身は保存されているので，式（6-18）は，エネルギー保存に関する事実としてつねに成立している．

$$Q = m_a c_a (T_2 - T_1) = -m_b c_b (T_4 - T_3) \qquad (6\text{-}18)$$

Where, the representative temperatures of high and low temperature substances can be noted by Eq. 6-19.

ここで，高温物体と低温物体の代表温度は式（6-19）のように記述することができる．

$$T_H = \frac{T_1 + T_2}{2}, \qquad T_L = \frac{T_3 + T_4}{2} \qquad (6\text{-}19)$$

Furthermore, Eq. 6-20 is known as an empirical fact of heat transport phenomena.

さらに，熱の移動現象についての経験上の事実として，式（6-20）が知られている．

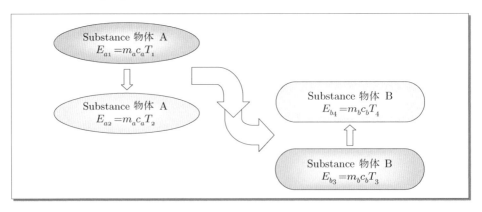

Fig.6-4   The second law of thermodynamics   熱力学の第二法則

$$T_H > T_L \qquad (6\text{-}20)$$

Using the above relationship, the following equation for entropy change can be obtained.

上記の関係を使えば，エントロピー変化についての次式を得ることができる．

$$\Delta S = Q\left(\frac{1}{T_L} - \frac{1}{T_H}\right) > 0, \quad \begin{array}{l}(\text{entropy change with heat transfer}\\ \quad \text{熱移動をともなうエントロピー変化})\end{array} \qquad (6\text{-}21)$$

This equation, which suggests that entropy change coupled with heat transport is always positive, is the formulated expression of the second law. Eq.6-22 ～ Eq.6-27 are the fundamental thermodynamic formulas related to entropy change.

この式は，熱の移動にともなうエントロピーの変化がつねに正の値となることを示していて，これが第二法則の定式化された表現である．式（6-22）から式（6-27）はエントロピーの変化に関連する熱力学的基礎式である．

$$ds = \frac{dq}{T}\left[\mathrm{J/(kg\cdot K)}\right] \qquad (6\text{-}22)$$

$$dS = \frac{dQ}{T}\left[\mathrm{J/K}\right] \qquad (6\text{-}23)$$

$$Tds = dq = du + Pdv \qquad (6\text{-}24)$$

$$TdS = dQ = dU + PdV \qquad (6\text{-}25)$$

$$Tds = dq = dh - vdP \qquad (6\text{-}26)$$

$$TdS = dQ = dH - VdP \qquad (6\text{-}27)$$

6.4 Free Energy

6.4 自由エネルギー

Consider a thermodynamic system composed of a high temperature heat source of temperature T_0 and a substance A of temperature T. Heat capacities of both heat source and substance A are assumed to be large enough so that

温度 T_0 の高温熱源と温度 T の物体からなる系を検討する．熱源と物体 A の熱量容は十分大きく，熱量 Q の移動の間に温度変化は起きないとする．

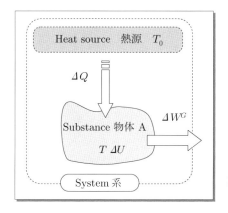

Fig.6-5　Work and internal energy
　　　　　仕事と内部エネルギー

no temperature changes result during the transportation of heat quantity Q.

Entropy change during the heat transportation process is simply expressed by Eq. 6-28.

熱の移動の間のエントロピーの変化は，式（6-28）のように問題なく表現できる．

$$\Delta S_0 = \frac{\Delta Q}{T_0}, \qquad \Delta S = \frac{\Delta Q}{T} \qquad (6\text{-}28)$$

During the process, substance A consumes part of the heat energy received from the heat source and does work. Then, the energy balance is noted by Eq. 6-29.

この熱の移動の間に，物体 A は熱源から受け取った熱エネルギーの一部を消費して仕事を行う．したがって，エネルギーバランスは，式（6-29）となる．

$$T \cdot \Delta S = \Delta U + \Delta W^G \qquad (6\text{-}29)$$

From the second law of thermodynamics, the temperature of heat source is higher than that of substance A, and the following relationship can be obtained.

熱力学の第二法則より，熱源の温度は物体 A より高いことになり，その結果，つぎの式が導かれる．

$$-\Delta U + T_0\ \Delta S \geq -\Delta U + T\ \Delta S = \Delta W^G \qquad (6\text{-}30)$$

Using the relationship given by Eq. 6-29, and assuming a quasi-equilibrium process, the following relationships can be obtained.

式（6-29）の関係と，この熱の移動過程についての準静的仮定を用いれば，以下のような関係を導くことができる．

$$\Delta(TS) = T \cdot \Delta S + S \cdot \Delta T \qquad (6\text{-}31)$$

$$T \cong T_0 \ \text{(quasi-thermal equilibrium 準熱平衡)} \Rightarrow \ \Delta(TS) \cong T\ \Delta S \cong T_0\ \Delta S \qquad (6\text{-}32)$$

$$-\Delta U + \Delta(TS) \geq \Delta W^G \qquad (6\text{-}33)$$

$$-\Delta(U - TS) \geq \Delta W^G \qquad (6\text{-}34)$$

Using the definition of Helmholtz's free energy as indicated in Eq. 6-35, Eq. 6-34 can be noted by Eq. 6-36.

式（6-35）に示すヘルムホルツの自由エネルギーの定義を用いれば，式（6-34）は式（6-36）のように記述することができる．

$$F \equiv U - TS \ \text{（Helmholtz'z free energy ヘルムホルツの自由エネルギー）} \qquad (6\text{-}35)$$

$$-\Delta F \geq \Delta W^G \qquad (6\text{-}36)$$

Equation 6-36 means that the decrease of free energy is always larger than the work. In other words, we cannot expect work to be larger than the decrease of free energy.

The following are fundamental equations related to Helmholtz's and Gibbs' free energy. Both are thermodynamic quantities of state and are effective in identifying the maximum work at constant temperature condition.

式（6-36）は，自由エネルギーの減少は仕事よりいつも大きいこと，言い換えれば，自由エネルギーの減少以上の仕事は期待できないことを意味している．

以下の式は，ヘルムホルツの自由エネルギーおよびギブスの自由エネルギーに関する基礎式である．両者のエネルギーとも等温条件下での最大仕事を明確に示すことに有効な熱力学的状態量である．

$$f = u - Ts \quad \text{(definition of Helmholtz's free energy [J/kg])} \tag{6-37}$$

$$F = U - TS \quad \text{(definition of Helmholtz's free energy [J])} \tag{6-38}$$

$$df = du - d(Ts) = du - Tds - sdT \tag{6-39}$$

$$Tds = dq = du + Pdv \tag{6-40}$$

$$df = -Pdv - sdT \tag{6-41}$$

$$g = h - Ts \quad \text{(definition of Gibbs' free energy [J/kg])} \tag{6-42}$$

$$G = H - TS \quad \text{(definition of Gibbs' free energy [J])} \tag{6-43}$$

$$dg = dh - d(Ts) = dh - Tds - sdT \tag{6-44}$$

$$Tds = dq = dh - vdP \tag{6-45}$$

$$dg = vdP - sdT \tag{6-46}$$

From the definition of the energy, it is clear that Helmholtz's free energy gives the maximum available energy in the internal energy, where as, Gibbs' free energy give the maxmum available enthalpy of flowning media.

このような自由エネルギーの定義から，ヘルムホルツの自由エネルギーは，内部エネルギーのうちの利用可能な最大エネルギーを，また，ギブスの自由エネルギーは流動する熱媒体の利用可能な最大エンタルピーを与えている.

Problems

問題

△6-1

Heat of 120 kJ is transferred from a heat source at 1 000 K to a working fluid at 300 K. If the temperatures of the heat source and the working fluid do not change in this process, determine the entropy loss of heat source and the entropy gain of working fluid. Does the total entropy change during the process increase or decrease?

△6-1

1 000 K の熱源から 300 K の作動流体に 120 kJ の熱が移動した．この過程において熱源が失ったエントロピーと作動流体が得たエントロピーを求めなさい．この過程で全体のエントロピーは増加するか，減少するか．

△6-2

Derive Eq. 6-49 from Eqs. 6-47 and 6-48.

△6-2

式（6-47）と式（6-48）から式（6-49）を導出しなさい.

$$TdS = dU + PdV \tag{6-47}$$

$$H = U + PV \tag{6-48}$$

$$TdS = dH - VdP \tag{6-49}$$

△6-3

Derive Eq. 6-50.

△6-3

式（6-50）を導出しなさい.

$$G = F + PV \tag{6-50}$$

Chapter 7
Chart of Thermodynamic Processes
熱力学的変化過程の線図

7.1 *P-v* Chart

Consider a general relationship between thermodynamic expansion process and work using the *P-v* chart shown in Fig.7-1. As explained in Chapter 4, expansion work produced by a gas expansion process from state 1 to state 2 is given by Eq. 7-1.

$$W_{1-2} = \int_1^2 Pdv \quad \text{(Expansion work 膨張仕事)} \tag{7-1}$$

On the other hand, technical work produced by this process is given by Eq. 7-2.

$$L_{1-2} = -\int_1^2 vdP \quad \text{(Technical work 工業仕事)} \tag{7-2}$$

Both of these works can be expressed with the areas such as $1\text{-}2\text{-}v_2\text{-}v_1$, and $1\text{-}2\text{-}P_2\text{-}P_1$, respectively. Thus, a *P-v* chart is useful to know the work corresponding to a process.

As for a constant pressure expansion process, the expansion work is given in Fig.7-2 and Eq. 7-3, but no technical work is done during the

7.1 *P-v* 線図

図 7-1 に示した *P-v* 線図を用いて，熱力学的な膨張過程と仕事の間の一般的な関係を検討する．第 4 章で説明したとおり，状態 1 から状態 2 へのガスの膨張過程によって生じる膨張仕事は，式（7-1）で与えられている．

一方，この過程で生じる工業仕事は式（7-2）で与えられる．

この二つの仕事は，面積 $1\text{-}2\text{-}v_2\text{-}v_1$ と面積 $1\text{-}2\text{-}P_2\text{-}P_1$ でそれぞれ与えられているので，*P-v* 線図は，変化過程に対応する仕事を知るのに有効である．

等圧膨張過程に関しては，膨張仕事が図 7-2 と式（7-3）で与えられているが，この変化からは工業仕事が生じない．逆に，式（7-4）

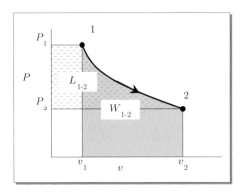

Fig.7-1 *P-v* chart and work
P-v 線図と仕事

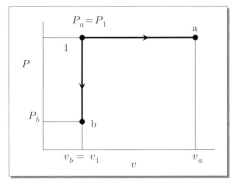

Fig.7-2 Thermodynamic process and *P-v* chart
熱力学的過程と *P-v* 線図

process. On the contrary, as indicated in Eq. 7-4, no expansion work is done during a constant volume pressure reduction process; only technical work is produced.

で示されているとおり，容積不変の圧力減少過程では膨張仕事は行われず，工業仕事のみが現れる．

$$\text{Process 1-a} \quad (P = \text{const.}) \qquad \left.\begin{array}{l} W = P_1(V_a - V_1) \\ L = 0 \end{array}\right\} \tag{7-3}$$

$$\text{Process 1-b} \quad (v = \text{const.}) \qquad \left.\begin{array}{l} W = 0 \\ L = -v_1(P_b - P_1) = v_1(P_1 - P_b) \end{array}\right\} \tag{7-4}$$

7.2 *T-s* Chart

7.2 *T-s* 線図

The general relationship between a thermodynamic process and the quantity of heat is clearly shown in a T-s chart as illustrated in Fig.7-3. According to the definition of entropy, the quantity of heat received during a thermodynamic process from state 1 to state 2 is obtained by an integration equation of Eq. 7-7.

熱力学的過程と熱量の一般的関係は，図7-3 に示した T-s 線図の中にわかりやすく表されている．エントロピーの定義に従えば，状態 1 から状態 2 への熱力学的過程の間に受け取る熱量は，式 (7-7) の積分によって与えられている．

$$ds = \frac{dq}{T} \tag{7-5}$$

$$dq = Tds \tag{7-6}$$

$$Q_{1-2} = \int_1^2 Tds \tag{7-7}$$

We can notice that the quantity of heat received during the process corresponds to the area of 1-2-s_2-s_1. As for an isothermal process (Fig.7-4), the integration operation of Eq. 7-7

この熱力学的過程の間に受け取った熱量は，面積 1-2-s_2-s_1 で表されている．等温度下における熱力学的過程（図 7-4）では，式 (7-7) の積分を容易に実行することができ，式 (7-8)

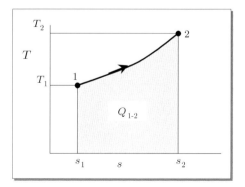

Fig.7-3 T-s chart and quantity of heat
T-s 線図と熱量

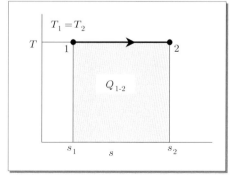

Fig.7-4 Isothermal process and T-s chart
等温過程と T-s 線図

can be done easily and Eq. 7-8 is obtained.   が得られる.

$$Q_{1\text{-}2} = T_1\left(s_2 - s_1\right) \qquad \text{(Process 1-2} \quad (T = \text{const.))} \tag{7-8}$$

7.3 Specific Heat and Entropy Change | ## 7.3 比熱とエントロピー変化

Here, we consider the relationship between specific heat and entropy. From the general equation of thermodynamic state shown in Eq. 7-9, and the definition of specific heat, the general specific heat indicated by Eq. 7-10 can be derived as:

ここで,比熱とエントロピーの関係を検討する.式 (7-9) に示されている一般的な状態方程式と比熱の定義より,式 (7-10) に示されている一般的な比熱が導出されている.

$$f(P,v,T)=0 \tag{7-9}$$

$$c = \frac{dq}{dT} = \frac{Tds}{dT} \tag{7-10}$$

Thus, the specific heats of constant volume and constant pressure can be derived respectively as follows:

したがって,等容比熱と等圧比熱はそれぞれつぎのように導出することができる.

$$dq = du + Pdv \tag{7-11}$$

$$\frac{dq}{dT} = \frac{du}{dT} + P\frac{dv}{dT} \tag{7-12}$$

$$du = \left(\frac{\partial u}{\partial T}\right)_v dT + \left(\frac{\partial u}{\partial v}\right)_T dv \tag{7-13}$$

$$c = \left(\frac{\partial u}{\partial T}\right)_v + \left[\left(\frac{\partial u}{\partial v}\right)_T + P\right]\frac{dv}{dT} \tag{7-14}$$

$$c_v = T\left(\frac{\partial s}{\partial T}\right)_v = \left(\frac{\partial q}{\partial T}\right)_v = \left(\frac{\partial u}{\partial T}\right)_v \tag{7-15}$$

$$dq = dh - vdP \tag{7-16}$$

$$\frac{dq}{dT} = \frac{dh}{dT} - v\frac{dP}{dT} \tag{7-17}$$

$$dh = \left(\frac{\partial h}{\partial T}\right)_P dT + \left(\frac{\partial h}{\partial P}\right)_T dP \tag{7-18}$$

$$c = \left(\frac{\partial h}{\partial T}\right)_P + \left[\left(\frac{\partial h}{\partial P}\right)_T - v\right]\frac{dP}{dT} \tag{7-19}$$

$$c_P = T\left(\frac{\partial s}{\partial T}\right)_P = \left(\frac{\partial q}{\partial T}\right)_P = \left(\frac{\partial h}{\partial T}\right)_P \tag{7-20}$$

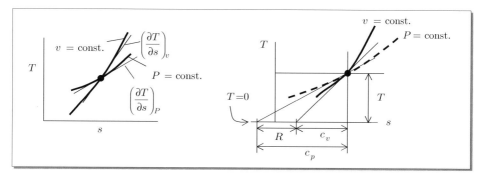

Fig.7-5　T-s chart and specific heat　T-s 線図と比熱

According to a geometrical analysis of the process on a T-s chart, the specific heat and the gradient of the process line shown in Fig.7-5 have the following relationship:

T-s 線図上の変化過程に対する幾何学上の解析によれば，図 7-5 に示した変化過程線の勾配と比熱は，以下のような関係を有している．

$$\left(\frac{\partial T}{\partial s}\right)_v = \frac{T}{c_v} \tag{7-21}$$

$$\left(\frac{\partial T}{\partial s}\right)_P = \frac{T}{c_p} \tag{7-22}$$

$$R = c_p - c_v = T\left(\frac{\partial s}{\partial T}\right)_P - T\left(\frac{\partial s}{\partial T}\right)_v \tag{7-23}$$

7.4 Other Useful Charts for Thermodynamics

7.4　熱力学上の他の有用な線図

The h-s chart is useful tool to determine the enthalpy change during an adiabatic process. In the adiabatic process, no quantity of heat is added to or released from the system. Considering the definition of entropy, we can see that no entropy change is produced in the adiabatic process. As for an adiabatic process shown in Fig.7-6, an increase in enthalpy is equivalent to a negative technical work as indicated by Eq. 7-25.

h-s 線図は断熱過程におけるエンタルピー変化を知るのに有効である．断熱変化では，対象としている系に熱が加えられることも，また，その系から熱が取り去られることもない．エントロピーの定義より，断熱過程ではエントロピー変化が生じないことは明らかである．図 7-6 の断熱過程では，式 (7-25) に示したとおり，エンタルピーの増加は負の工業仕事に相当する．

$$dq = dh - vdP \tag{7-24}$$

Process 1-a $(s = \text{const.})$

$$h_{1-a} = \int_1^a dq + \int_1^a vdP = \int_1^a vdP = -L_{1-a} = h_2 - h_1 \tag{7-25}$$

When a pressure is kept constant during a

熱力学的過程の際に圧力が一定であれば，

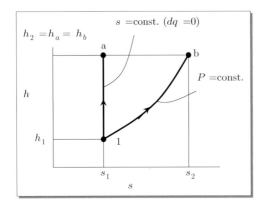

Fig.7-6 h-s chart (Mollier chart)
　　　　h-s 線図 （モリエ線図）

process, Eq. 7-26 shows that enthalpy change is equivalent to a quantity of heat that is received during the process.

式（7-26）がエンタルピーの変化はその間に受け取った熱量に相当することを示している.

Process 1-b $(p = \text{const.})$

$$h_{1-b} = \int_1^b dq + \int_1^b vdP = \int_1^b dq = q_{1-b} = h_2 - h_1 \qquad (7\text{-}26)$$

The relationship between internal energy and entropy is also important for thermodynamics of a gas-filled vessel. From the analysis of a u-s chart in Fig.7-7, we can determine the change of internal energy for various processes.

内部エネルギーとエントロピーの関係も, 容器に満たされた気体の熱力学にとって重要である. 図7-7の u-s 線図での解析から, 種々の熱力学的過程における内部エネルギーを知ることができる.

$$dq = du + Pdv \qquad (7\text{-}27)$$

Process 1-a $(s = \text{const.})$

$$u_{1-a} = \int_1^a dq - \int_1^a vdP = -\int_1^a Pdv = -W_{1-a} = u_2 - u_1 \qquad (7\text{-}28)$$

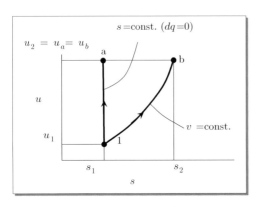

Fig.7-7 u-s chart u-s 線図

Process 1-b $(v = \text{const.})$

$$u_{1-b} = \int_1^b dq - \int_1^b Pdv = \int_1^b dq = q_{1-b} = u_2 - u_1 \qquad (7\text{-}29)$$

Problems	**問題**

△7-1

The thermodynamic charge of a gas on the $T\text{-}S$ chart shown in Fig.7-8. Determine the amount of heat which the gas receives in this process.

△7-1

あるガスの熱力学的変化が $T\text{-}S$ 線図で図 7-8 のように表されている．この変化の過程において ガスが受け取った熱量を求めなさい．

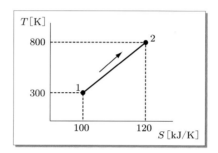

Fig.7-8 $T\text{-}S$ chart for Problem 7-1　問題 7-1 に関する $T\text{-}S$ 線図

△7-2

1 kg of water at 5 °C is heated. The increase in entropy of water is 1.2 kJ/K in this heating process. If the specific heat of water is 4.19 kJ/(kg·K) and is independent of temperature, determine the temperature of water after heating.

△7-2

5 °C，1 kg の水が加熱される．この加熱過程で 水のエントロピーが 1.2 kJ/K 増加した．水の比熱 が 4.19 kJ/(kg·K) で温度に依存しないとするとき， 加熱後の水の温度を求めなさい．

Chapter 8
Thermodynamic Process of Ideal Gas
理想気体の状態変化

8.1 Isothermal Change

8.1 等温変化

Since the assumption of an ideal gas is usually applied to a gaseous substance, the thermodynamic process of an ideal gas is very important for the fundamentals of thermodynamics. As for an ideal gas of which the equation of state is noted by Eq.8-1, the internal energy, enthalpy and quantity of heat during a process can be given by Eq.8-2 and Eq.8-3.

理想気体という仮定が気体状態の物質に通常適用されるため，理想気体の熱力学的な変化過程は熱力学の基礎としてたいへん重要である．状態方程式が式（8-1）で与えられる理想気体については，変化過程における内部エネルギー，エンタルピー，熱量が式（8-2）と式（8-3）で与えられる．

$$Pv = RT \tag{8-1}$$
$$du = c_v dT, \quad dh = c_p dT, \quad dq = Tds \tag{8-2}$$
$$dq = c_v dT + Pdv = c_p dT - vdP \tag{8-3}$$

Here consider an isothermal process as shown in Fig.8-1. During an isothermal change, the temperature is kept constant and the equation of the process can be given by Eq.8-4.

ここで，図 8-1 のような等温変化を考える．等温変化の間では，温度は一定であり，その結果として，変化過程の方程式は式（8-4）で与えられる．

$$T = \text{constant} \quad \Rightarrow \quad dT = 0, \quad Pv = \text{constant} \tag{8-4}$$

Using Eq.8-3, the quantity of heat gained by the ideal gas during a change from state 1 to state 2 can be given by Eq.8-5.

式（8-3）を用いると，状態 1 から状態 2 への変化の間に気体が受け取る熱量は，式（8-5）で与えられることになる．

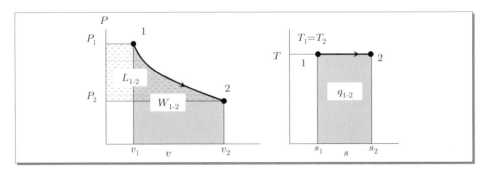

Fig.8-1 Isothermal change　等温変化

$$q_{1-2} = \int_1^2 P dv = -\int_1^2 v dP = W_{1-2} = L_{1-2} \tag{8-5}$$

We can find out that the expansion work in an isothermal change is equivalent to the technical work. Furthermore, using the equation of state and assuming a condition of no temperature change, the quantity of heat during the isothermal change can be given by Eq.8-6 or Eq.8-7.

また，等温変化における膨張仕事は，そこでの工業仕事と等しいことがわかる．さらに，理想気体の方程式と温度変化がないことから，等温変化過程での熱量は，式（8-6）または式（8-7）で与えられる．

$$q_{1-2} = \int_1^2 P dv = \int_1^2 \frac{RT}{v} dv = RT \int_1^2 \frac{dv}{v} = RT \ln \frac{v_2}{v_1} \tag{8-6}$$

$$q_{1-2} = RT \ln \frac{v_2}{v_1} = P_1 v_1 \ln \frac{v_2}{v_1} = P_1 v_1 \ln \frac{P_1}{P_2} \tag{8-7}$$

From the above equations, the entropy change can be derived as shown in the equation below.

これらの式から，式（8-8）で示されるエントロピー変化が導出される．

$$s_{1-2} = \int_1^2 ds = \int_1^2 \frac{dq}{T} = \frac{1}{T} \int_1^2 dq = R \ln \frac{v_2}{v_1} \tag{8-8}$$

| ## 8.2　Isobaric Change | ## 8.2　等圧変化 |

When the pressure is kept constant during a thermodynamic change (Fig.8-2), the equation of the process becomes Eq.8-9. This process is called with isobaric change or constant pressure change.

熱力学的な変化において圧力が一定であるならば（図 8-2），この変化の方程式は式（8-9）となる．この過程は等圧変化または定圧変化とよばれている．

$$P = \text{constant} \quad \Rightarrow \quad dP = 0, \quad \frac{v}{T} = \frac{R}{P} = \text{constant} \tag{8-9}$$

Using Eqs.8-1〜8-3 and the definition of entropy, we can obtain the following equations:

さらに，式（8-1）〜（8-3），およびエントロピーの定義により，以下の式が得られる．

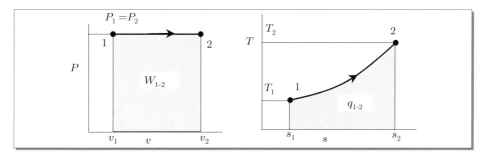

Fig.8-2　Isobaric（constant pressure）change　等圧（定圧）変化

$$W_{1-2} = \int_1^2 Pdv = P\int_1^2 dv = P(v_2 - v_1) = R(T_2 - T_1) \tag{8-10}$$

$$L_{1-2} = 0 \tag{8-11}$$

$$q_{1-2} = \int_1^2 c_p dT = c_p(T_2 - T_1) = h_2 - h_1 \tag{8-12}$$

$$s_{1-2} = \int_1^2 ds = \int_1^2 \frac{dq}{T} = \int_1^2 \frac{c_p dT}{T} = c_p \ln \frac{T_2}{T_1} \tag{8-13}$$

8.3 Isochoric Change

The equation of a process for an isochoric (constant volume) change (Fig.8-3) of an ideal gas is given by Eq.8-14.

8.3 等容変化

理想気体に対する等容(定容)変化の過程 (図 8-3) を表す方程式は,式 (8-14) で与えられる.

$$v = \text{constant} \;\Rightarrow\; dv = 0, \;\; \frac{P}{T} = \frac{R}{v} = \text{constant} \tag{8-14}$$

Since no volumetric displacement is produced during the isochoric change, no expansion work is obtained. However, technical work can be described by Eq.8-16. The quantity of heat and entropy change is given by Eqs.8-17 and 8-18.

等容変化では体積変化が起きないので,膨張仕事は生じない.しかし,工業仕事は式 (8-16)で記述されている.熱量とエントロピー変化は,式 (8-17) と式 (8-18) で与えられる.

$$W_{1-2} = 0 \tag{8-15}$$

$$L_{1-2} = -\int_1^2 vdP = -v\int_1^2 dP = v(P_1 - P_2) = R(T_1 - T_2) \tag{8-16}$$

$$q_{1-2} = \int_1^2 c_v dT = c_v(T_2 - T_1) = u_2 - u_1 \tag{8-17}$$

$$s_{1-2} = \int_1^2 ds = \int_1^2 \frac{dq}{T} = \int_1^2 \frac{c_v dT}{T} = c_v \ln \frac{T_2}{T_1} \tag{8-18}$$

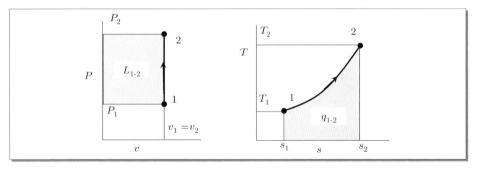

Fig.8-3   Isochoric (constant volume) change   等容(定容)変化

8.4 Adiabatic Change

From the entropy definition noted in Eq.6-22, an adiabatic change means a thermodynamic change of constant entropy. As for the quantity of heat during the adiabatic change, the following relationships can then be obtained.

8.4 断熱変化

式 (6-22) に記述したエントロピーの定義より，断熱変化はエントロピー一定の熱力学的変化を意味している．したがって，断熱変化の際の熱量については，以下の関係式が得られる．

$$dq = 0 \quad \Rightarrow \quad ds = 0, \quad s = \text{constant} \tag{8-19}$$

$$dq = c_v dT + Pdv = 0 \tag{8-20}$$

$$dq = c_p dT - vdP = 0 \tag{8-21}$$

From the total differential of the equation of state and Eq.8-20, we can derive the following equations:

理想気体の状態方程式の全微分と式 (8-20) から，以下の式を導くことができる．

$$Pdv + vdP = RdT = -\frac{RPdv}{c_v} \tag{8-22}$$

$$(c_v + R)Pdv + c_v vdP = 0 \tag{8-23}$$

$$c_p \frac{dv}{v} + c_v \frac{dP}{P} = 0 \tag{8-24}$$

The differential equation of Eq.8-24 can be solved and its solution is written in the form: Eq.8-25.

微分方程式 (8-24) は解くことができ，式 (8-25) が解として得られる．

$$\kappa \ln v + \ln P = C \tag{8-25}$$

The result shown in Eq.8-26 is the equation for an adiabatic process.

式 (8-26) で示される結果が，断熱変化過程に対する状態方程式となる．

$$Pv^\kappa = \text{constant} \tag{8-26}$$

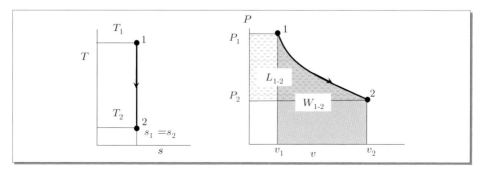

Fig.8-4　Adiabatic change　断熱変化

Moreover, Eq.8-27 and Eq.8-28 are the other forms of the equation of state for an adiabatic change.

さらに，断熱変化についての別な方程式の形式として，式（8-27）と式（8-28）が得られる．

$$Tv^{\kappa-1} = \text{constant} \tag{8-27}$$

$$\frac{P^{\frac{\kappa-1}{\kappa}}}{T} = \text{constant} \tag{8-28}$$

Considering the internal energy during an adiabatic change, the expansion work can be given by Eq.8-30.

断熱変化の際の内部エネルギーを検討することで，膨張仕事が式（8-30）のように得られる．

$$dq = Tds = du + Pdv = 0 \tag{8-29}$$

$$W_{1-2} = \int_1^2 Pdv = -(u_2 - u_1) = -c_v(T_2 - T_1) = \frac{1}{\kappa-1}R(T_1 - T_2) = \frac{1}{\kappa-1}(P_1v_1 - P_2v_2) \tag{8-30}$$

Using the equation of enthalpy, the technical work can be given as follows:

エンタルピーの式を使えば，工業仕事は以下のようになる．

$$dq = Tds = dh - vdP = 0 \tag{8-31}$$

$$L_{1-2} = -\int_1^2 vdP = -(h_2 - h_1) = -c_p(T_2 - T_1) = \frac{\kappa}{\kappa-1}R(T_1 - T_2) = \frac{\kappa}{\kappa-1}(P_1v_1 - P_2v_2) \tag{8-32}$$

8.5 Polytropic Change

8.5 ポリトロープ変化

Consider a thermodynamic change coupled with a heat transport phenomena but is subjected to the process equation similar to the adiabatic change. This process can be assumed as shown in Eq.8-34 and is called a polytropic change.

熱の移動が行われ，かつ断熱変化と類似した方程式で変化過程が規定されている熱力学的変化を検討する．この変化過程は式（8-34）のように想定することができ，ポリトロープ変化とよばれている．

$$dq = Tds = du + Pdv \neq 0 \tag{8-33}$$

$$Pv^n = \text{constant} \tag{8-34}$$

A typical polytropic change on a P-v chart is shown in Fig.8-5.

図8-5の P-v 線図には，典型的なポリトロープ変化が示してある．

Using the definition of polytropic change (Eq.8-34) and a differential form of the equation of state (Eq.8-35), the following equations concerning to the polytropic change are obtained. Here $n > 1$ is assumed for simmplicity of analysis.

式（8-34）のポリトロープ変化の定義と，式（8-35）の状態方程式の微分形を用いれば，ポリトロープ変化に関するつぎの関係式が得られる．ここで，解析を簡単にするため $n > 1$ を想定しておく．

$$Pdv + vdP = RdT \tag{8-35}$$

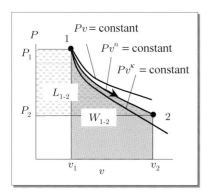

Fig.8-5 Polytropic change
ポリトロープ変化

$$Tv^{n-1} = \text{constant} \tag{8-36}$$

$$\frac{P^{\frac{n-1}{n}}}{T} = \text{constant} \tag{8-37}$$

$$nPv^{n-1}dv + v^n dP = 0 \tag{8-38}$$

These equations are similar to the equations of adiabatic change except for the difference of κ and n. As for the works during the change, they are also similar to the works during an adiabatic change.

これらの関係式は，κ と n の違いを除けば断熱変化の関係式と類似している．変化の間の仕事に関しても，断熱変化の場合の仕事に類似したものになる．

$$W_{1\text{-}2} = \int_1^2 P dv = -\int_1^2 \frac{1}{n-1} R dT = \frac{1}{n-1} R(T_1 - T_2) = \frac{1}{n-1}(P_1 v_1 - P_2 v_2) \tag{8-39}$$

$$L_{1\text{-}2} = -\int_1^2 v dP = -\int_1^2 \frac{n}{n-1} R dT = \frac{n}{n-1} R(T_1 - T_2) = \frac{n}{n-1}(P_1 v_1 - P_2 v_2) \tag{8-40}$$

The quantity of heat can be derived from either the internal energy equation or the enthalpy equation.

熱量については，内部エネルギーの式や，エンタルピーの式のいずれからでも導出することができる．

$$q_{1\text{-}2} = \int_1^2 du + \int_1^2 P dv = c_v(T_2 - T_1) + W_{1\text{-}2} = \left(c_v - \frac{c_v(\kappa-1)}{n-1} \right)(T_2 - T_1)$$

$$= \left(\frac{n-k}{n-1} \right) c_v(T_2 - T_1) \tag{8-41}$$

$$q_{1\text{-}2} = \int_1^2 dh - \int_1^2 v dP = c_p(T_2 - T_1) + L_{1\text{-}2} = \left(c_p - \frac{nc_p\dfrac{\kappa-1}{\kappa}}{n-1} \right)(T_2 - T_1)$$

$$= \frac{1}{\kappa}\left(\frac{n-k}{n-1} \right) c_p(T_2 - T_1) \tag{8-42}$$

Here the specific heats for polytropic change

ここで，ポリトロープ変化に対する比熱は，

are defined by Eq.8-43.   式（8-43）で定義されている.

[Polytropic specific heat　ポリトロープ比熱]

$$c_{nv} = \frac{n-\kappa}{n-1}c_v, \quad c_{np} = \frac{n-\kappa}{n-1}c_p \qquad (8\text{-}43)$$

Entropy change during a polytropic process is   ポリトロープ変化にともなうエントロピー
derived as follows: 変化は，つぎのようになる.

$$s_{1\text{-}2} = \int_1^2 \frac{dq}{T} = \int_1^2 \frac{n-\kappa}{n-1}c_v \frac{dT}{T} = \frac{n-\kappa}{n-1}c_v \ln\frac{T_2}{T_1} \qquad (8\text{-}44)$$

Polytropic change is generally characterized by "n" which is called the polytropic exponent. Figure 8-6 shows various thermodynamic changes expressed by the process equations of polytropic change. A process that has polytropic exponent of $n = 0$ corresponds to a constant pressure process. Processes of $n = 1$ and $n = \kappa$ corresponds to an isothermal process and an adiabatic process, respectively. A constant volume process is indicated by a polytorpic change of $n = \infty$.

ポリトロープ変化は，ポリトロープ指数とよばれる n によって，一般に特徴づけられている. 図8-6は，ポリトロープ変化に対応する変化過程の方程式で表現されている種々の熱力学的変化である. ポリトロープ指数が $n = 0$ である場合には，等圧変化過程となる. $n = 1$ と $n = \kappa$ の変化は，それぞれ等温変化と断熱変化に対応している. また，等容変化は，$n = \infty$ のポリトロープ変化となる.

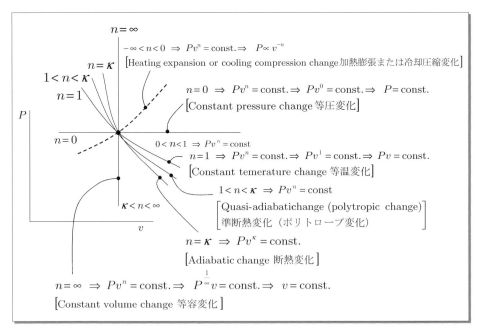

Fig.8-6 Polytropic changes for various n　種々の n に対するポリトロープ変化

Problems

問題

△8-1

An ideal gas of 0.2 m³ at 5 MPa expands to 4 m³ under a constant temperature of 300 K. Determine (1) the final pressure of the gas, (2) the work done during this process, and (c) the entropy change in the process.

△8-2

3 kg of an ideal gas at 300 K is heated to 400 K under a constant pressure of 0.1 MPa. If the specific heat at constant pressure is 1.005 kJ/(kg·K) and the gas constant is 0.2871 kJ/(kg·K), determine (1) the final volume of the gas, (2) the expansion work done during this process, and (3) the entropy change in the process.

△8-3

3 kg of an ideal gas at 250 K is heated to 450 K under a constant volume of 0.1 m³. If the specific heat at constant volume is 0.716 kJ/(kg·K) and the gas constant is 0.2871 kJ/(kg·K), determine (1) the final pressure of the gas, (2) the total heat transferred to the gas, and (3) the change in entropy in this process.

△8-4

A frictionless piston and cylinder device (see Fig.8-1) initially contains 1 m³ of an ideal gas at 0.2 MPa and 300 K. Now, the gas is compressed to 0.2 m³ without heat transfer. If the specific heat at constant pressure is 1.005 kJ/(kg·K) and the gas constant is 0.2871 kJ/(kg·K), determine (1) the final pressure of the gas, (2) the final temperature of the gas, and (3) the work done during this process.

△8-5

5 m³ of an ideal gas at 0.1 MPa is compressed to 15 MPa. The final volume of the gas is 0.078 m³. Determine the polytropic exponent n for this process.

△8-1

5 MPa で 0.2 m³ の理想気体が 300 K の一定温度下で 4 m³ まで膨張する．このとき，(1) 最終的な気体の圧力，(2) この過程でなされた仕事，(3) この過程におけるエントロピー変化をそれぞれ求めなさい．

△8-2

300 K の理想気体 3 kg を 0.1 MPa の一定圧力下で 400 K まで加熱する．定圧比熱を 1.005 kJ/(kg·K)，ガス定数を 0.2871 kJ/(kg·K) とするとき，(1) 最終的な気体の体積，(2) この過程でなされた膨張仕事，(3) この過程におけるエントロピー変化をそれぞれ求めなさい．

△8-3

250 K の理想気体 3 kg を 0.1 m³ の一定容積下で 450 K まで加熱した．定容比熱を 0.716 kJ/(kg·K)，ガス定数を 0.2871 kJ/(kg·K) とするとき，(1) 最終的な気体の圧力，(2) この過程で気体に加えられた熱量，(3) この過程におけるエントロピー変化をそれぞれ求めなさい．

△8-4

摩擦のないピストンとシリンダからなる装置（図 8-7）の内部に，初期状態で 0.2 MPa，300 K の理想気体が 1 m³ 入っている．ここで，気体を熱の移動なしに 0.2 m³ まで圧縮する．定圧比熱を 1.005 kJ/(kg·K)，ガス定数を 0.2871 kJ/(kg·K) とするとき，(1) 最終的な気体の圧力，(2) 最終的な気体の温度，(3) この過程の間になされた仕事をそれぞれ求めなさい．

△8-5

0.1 MPa の理想気体 5 m³ を 15 MPa まで圧縮する．圧縮後の体積は 0.078 m³ である．この過程におけるポリトロープ指数 n を求めなさい．

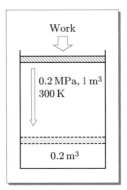

Fig.8-7   Schematic for Problem 8-4　問題 8-4 の説明図

Chapter 9
Reversible and Irreversible Changes
可逆変化と非可逆変化

9.1 Reversible and Irreversible Changes

Using the illustration in Fig.9-1, consider a general thermodynamic process from state 1 to state 2. During the process, a quantity of heat ΔQ is given to the working gas and work ΔW^G is done by the working gas. As for the system that includes a heat source, a working gas and work, the above process means that the quantity of heat released from the heat source is equivalent to a change of internal energy of the working gas and the external work done by the gas.

Then, consider the reverse process from state 2 to state 1. If the working gas is capable of

9.1 可逆変化と非可逆変化

図 9-1 の図解をもとに，状態 1 から状態 2 への一般的な熱力学的変化を検討する．変化の過程では，作動気体に対して熱量 ΔQ が与えられ，作動気体は仕事 ΔW^G を行う．熱源と作動気体と仕事を含む系全体でみれば，上記の変化過程は，熱源から放出された熱量が作動気体の内部エネルギーの増加と，気体による外部仕事に等しいことを意味している．

つぎに，状態 2 から状態 1 への逆変化過程を検討する．負の仕事 $-\Delta W^G$ を消費し，か

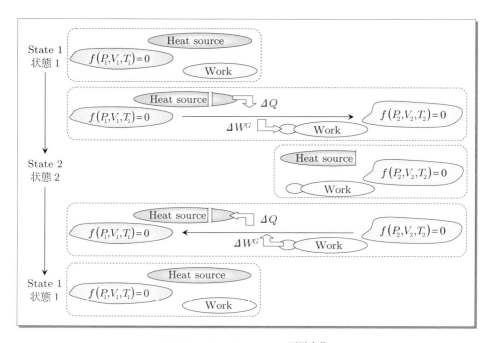

Fig.9-1 Reversible change 可逆変化

changing back to state 1 consuming a negative work $-\Delta W^{G}$ and returning the quantity of heat ΔQ, the thermodynamic change of this process is called a reversible change. A couple of forward and backward reversible processes are characterized by the fact that no traces remain after these forward and backward changes occur in a system.

On the contrary, as shown in Fig.9-2, if some part of the released quantity of heat leaks out of the system, the total energy change due to the change of internal energy of the working gas and the work done by the gas is less than the amount of the heat released from the heat source. Furthermore, during the backward change from state 2 to state 1, if some part of the work is not effectively used in the backward process of the gas, the heat source can not recover the released quantity of heat as a whole. The thermodynamic change explained here is called an irreversible change.

つ熱源に熱量 ΔQ を熱源に返しながら，作動気体を状態1にもどすことができたとすれば，その過程の熱力学的変化は可逆変化とよばれる．順方向および逆方向の一組の可逆変化は，熱力学的な系内において，これらの順方向と逆方向の変化のあとに何の変化の痕跡も残さないことで特徴づけられる．

上記とは異なり，図9-2に示すとおり，もし放出された熱量の一部が対象としている系の外に逃げるとすれば，作動気体の内部エネルギーの変化と，気体がなした仕事の和としての全体のエネルギー変化は，熱源から放出された熱量より小さくなる．さらに，状態2から状態1への逆変化の際に，仕事のある部分が対象とする気体が状態1にもどる過程に有効に使われていないならば，全体として熱源は最初に放出した熱を取りもどすことができない．ここで説明した熱力学的変化は，非可逆変化とよばれている．

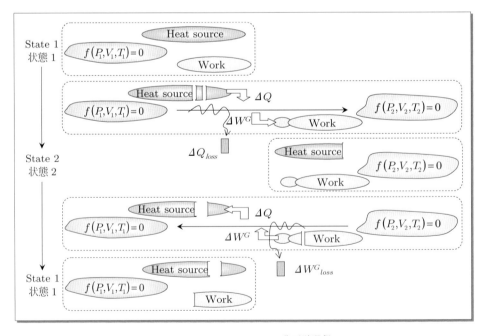

Fig.9-2   Irreversible change   非可逆過程

9.2 Forward and Backward Changes

9.2 順方向変化と逆方向変化

Here a working gas of 1 kg is subjected to a thermodynamic change as expressed in Eq.9-2 and this is assumed as a reversible change of an ideal gas. The process of change, where positive work is done, should be called as a forward change and the reverse process should be a backward change.

式（9-2）に示される熱力学的変化を行っている 1 kg の作動気体について検討する．さらに，この熱力学的変化は理想気体の可逆変化であるとする．正の仕事が行われる変化過程を順方向変化，また，その逆の過程を逆方向変化とするのが理に適っている．

$$Pv = RT \tag{9-1}$$

$$P = f_{route}(v, T) \tag{9-2}$$

According to the forward process shown in Fig.9-3, a quantity of heat during a change from state 1 to state 2 can be noted by Eq.9-3.

図 9-3 に示した順方向変化では，状態 1 から状態 2 への変化の際の熱量は式（9-3）となる．

Change from 1 to 2

$$q_{1\text{-}2} = \int_1^2 c_v dT + \int_1^2 P dv = c_v (T_2 - T_1) + W_{1\text{-}2} \tag{9-3}$$

Using the process equation and the quantity of heat during the change, the thermodynamic quantities at state 2 are obtained as follows:

状態変化の方程式と変化の際の熱量から，状態 2 における熱力学的諸量は，以下のようになる．

$$P_2 = f_{route(1\text{-}2)}(v_2, T_2) \tag{9-4}$$

$$T_2 = T_1 + \frac{q_{1\text{-}2} - W_{1\text{-}2}}{c_v} \tag{9-5}$$

$$v_2 = \frac{RT_2}{P_2} \tag{9-6}$$

As for the backward change, the negative work needed to return to state 1 is given by Eq.9-7.

逆方向の変化に関しては，状態 1 にもどるための負の仕事は式（9-7）で与えられる．

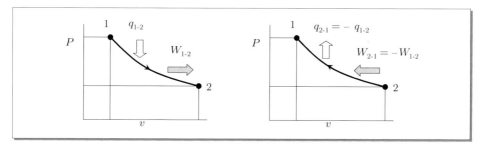

Fig.9-3　Forward and backward changes　順方向変化と逆方向変化

$$W_{2-1} = \int_2^1 Pdv = -\int_1^2 Pdv = -W_{1-2} \tag{9-7}$$

The quantity of heat during the change is given by Eq.9-8. Here, T' is a tentative temperature at the end of the backward change.

この変化の際の熱量は，式（9-8）で与えられている．ここで，T' は逆方向変化の終了状態における仮の温度とする．

Change from 2 to 1

$$q_{2-1} = \int_2^1 c_v dT + \int_2^1 Pdv = c_v(T_1' - T_2) + W_{2-1} \tag{9-8}$$

According to the reversible processes of forward and backward changes, the thermodynamic quantities at the end of the backward change should be consistent with the quantities of the initial state 1. Then, the following relationships are derived for forward and backward changes of a reversible process.

順方向変化と逆方向変化が可逆変化であることから，逆方向変化の最終状態での熱力学的諸量は，状態 1 の初期の諸量に一致するはずである．したがって，可逆変化過程での順方向変化と逆方向変化については，以下の関係式が成り立つ．

$$(P_1', v_1', T_1') = (P_1, v_1, T_1) \tag{9-9}$$

$$T_1' = T_2 + \frac{q_{2-1} - W_{2-1}}{c_v} = T_1 = T_2 - \frac{q_{1-2} - W_{1-2}}{c_v} \tag{9-10}$$

$$q_{2-1} - W_{2-1} = -q_{1-2} + W_{1-2} \tag{9-11}$$

$$q_{2-1} = -q_{1-2} \tag{9-12}$$

When the forward change includes an irreversible change, a part of the heat received is released as heat loss. The quantity of heat during an irreversible change from state 1 to state 2 is noted by Eq.9-13.

ここで，順方向変化に非可逆変化が含まれると，その結果として，作動気体が受け取った熱の一部分が熱損失として失われる．状態 1 から状態 2 への非可逆過程の場合の熱量は，式（9-13）のように記述される．

Change from 1 to 2

$$q_{1-2} = \int_1^2 c_v dT + \int_1^2 Pdv + q_{loss(1-2)} = c_v(T_2 - T_1) + W_{1-2} + q_{loss(1-2)} \tag{9-13}$$

Heat loss is always positive and Eq.9-14 shows the typical characteristics of an irreversible process.

ここで，熱損失はつねに正の値であるので，結果として式（9-14）が非可逆変化の特徴を与えることになる．

$$|q_{1-2}|_{irreversible} > |q_{1-2}|_{reversible} \tag{9-14}$$

When the backward change includes an irreversible change, some of the negative work received by the gas is not used to increase the internal energy of the gas. Thus, for an irreversible change, the following relationships

逆方向変化が非可逆変化を含む場合では，作動気体が受け取る負の仕事のある部分が作動気体自身の内部エネルギーの増加に使われない．その結果，非可逆変化について以下のような式を得ることができる．

can be derived:

Change from 2 to 1

$$q_{2-1} + W_{loss(2-1)} = \int_2^1 c_v dT + \int_2^1 P dv + W_{loss(2-1)} = c_v (T_1 - T_2) + W_{irreversible(2-1)} \qquad (9\text{-}15)$$

$$|W_{2-1}|_{irreversible} > |W_{2-1}|_{reversible} \qquad (9\text{-}16)$$

9.3 Heat Transfer Phenomena and Irreversible Process

9.3 熱伝達現象と非可逆変化

Using the illustration shown in Fig.9-4, consider the heating process of water in a kettle. One process is the direct heating of the kettle where all the quantity of heat produced from the flame is assumed to be effectively used to increase the internal energy of water in the kettle. In the other process, the quantity of heat from the flame is once converted to the internal energy of the working gas and potential energy of the weight.

Both of these energies are then transferred to the water in the kettle. The following assumptions are now used for analysis:

(1) Ideal gas assumption for working gas
(2) Quasi-equilibrium heat transfer
(3) No heat loss
(4) Constant pressure reversible change
(5) Quasi-equilibrium expansion process

やかんの中の水を温める過程を，図 9-4 に示したイラストを用いて検討する．一つの方法は，やかんを直接加熱する方法であり，炎によって生じたすべての熱はやかんの中の水の内部エネルギーを増加させるために効果的に使われていると想定する．別の方法として，炎からの熱量をいったん作動気体の内部エネルギーとおもりの位置エネルギーに変換する．

その両方のエネルギーは，その後，やかんの中の水に伝わる．ここで，解析のために以下の事項を想定する．

（1）作動気体は理想気体とする．
（2）熱伝達は準静的に行われるとする．
（3）熱損失がないとする．
（4）可逆等圧変化とする．
（5）準静的膨張過程とする．

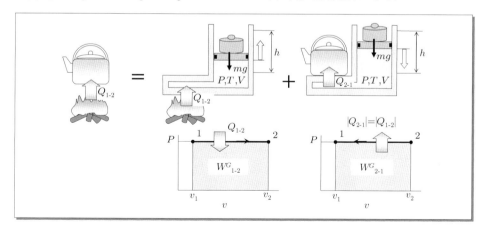

Fig.9-4 Heat transfer phenomena with reversible process of thermodynamic change
可逆的な熱力学的変化をともなう熱伝達現象

(6) Quasi-equilibrium compression process

During an expansion process, the quantity of heat transferred to the working gas is given by Eq.9-17.

(6) 準静的圧縮過程とする.

膨張過程において，作動気体に伝達された熱量は式（9-17）にて与えられる.

Change from 1 to 2

$$Q_{1-2} \Rightarrow G \int_1^2 c_v dT + \int_1^2 P dV = G c_v (T_2 - T_1) + W_{1-2}^G \Rightarrow G c_v (T_2 - T_1) + mg(h_2 - h_1) \quad (9\text{-}17)$$

Furthermore, during the compression process, this heat is transferred to the water in the kettle.

さらに，圧縮過程において，この熱はやかんの中の水に伝えられる.

Change from 2 to 1

$$G c_v (T_1 - T_2) + mg(h_1 - h_2) \Rightarrow G c_v (T_1 - T_2) + W_{2-1}^G$$
$$= G \int_2^1 c_v dT + \int_2^1 P dV \Rightarrow Q_{2-1} = -Q_{1-2} \quad (9\text{-}18)$$

Eq.9-18 indicates that the quantity of heat transferred to the water through reversible processes is absolutely equivalent to the quantity of heat that is transferred directly from the flame to the water.

On the other hand, when irreversible processes such as friction loss and heat loss are included

式（9-18）は，可逆過程を介して水に伝えられる熱量が，炎から水に直接伝えられる熱量に完全に一致することを示している.

ところが，摩擦損失や熱損失のような非可逆過程が，作動気体の膨張過程や圧縮過程の

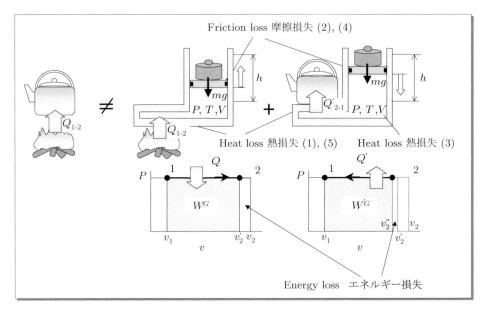

Friction loss 摩擦損失 (2), (4)
Heat loss 熱損失 (1), (5)   Heat loss 熱損失 (3)
Energy loss エネルギー損失

Fig.9-5   Heat transfer phenomena with irreversible process of thermodynamic change
非可逆的な熱力学的の変化をともなう熱伝達現象

in the expansion and compression processes of the working gas, the quantity of heat decreases step by step as shown in Fig.9-5 and Eq.9-19.

中に含まれている場合では,図 9-5 や式(9-19)で示すように,熱量はつぎつぎと変化の段階を経るごとに減少していく.

$$|Q_{1-2}| > |Q| > |W^G| > |W'^G| > |Q'| > |Q'_{2-1}| \qquad (9\text{-}19)$$
$$\uparrow \quad \uparrow \quad \uparrow \quad \uparrow \quad \uparrow$$
$$(1) \quad (2) \quad (3) \quad (4) \quad (5)$$

(1) Haet loss in heat transfer　熱伝達における熱損失

(2) Friction loss on piston movement　ピストンの運動にともなう摩擦損失

(3) Heat loss throgh container wall　容器の壁面からの熱損失

(4) Friction loss on piston movement　ピストンの運動にともなう摩擦損失

(5) Heat loss in heat transfer　熱伝達における熱損失

It means that the quantity of heat transfered through irreversible process is less then the heat supplied by high temperature heat source.

このことは,非可逆過程を経て伝達される熱量は,高温熱量が供給した熱量より少ないことを示している.

Problems

△9-1

2 kg of ideal gas at 300 K is heated to 500 K under constant pressure. The actual work done during this process is 70 kJ. If the gas constant is 0.1889 kJ/(kg·K), determine the energy loss during this process.

問題

△9-1

300 K の理想気体 2 kg が 500 K まで加熱される.この過程の間でなされた実際の仕事は 70 kJ である.ガス定数を 0.1889 kJ/(kg·K) とするとき,この過程の間のエネルギー損失を求めなさい.

Chapter 10
Mixing, Throttling and Filling
混合と絞りと充填

10.1 Mixing and Dalton's Law of Partial Pressure

10.1 混合とドルトンの分圧の法則

Consider a mixing problem shown in Fig.10-1. As for a simple case, the pressures and temperatures of gases to be mixed are assumed to be same and will be called the common pressure and temperature. However, each gas has its own the mass and volume, and their gas constants are different owing to the different characteristics of each gas. Furthermore, no addition of work and heat is assumed during mixing.

The gas labeled as i-th is subjected to its own equation of state and the relationships indicated in Eq.10-1 always exist among these gases.

図 10-1 に示す混合問題を検討する．簡単な場合として，混合すべきそれぞれの気体の圧力と温度は等しいとし，それを共通の圧力および温度とよぶことにする．一方，それぞれの気体は，気体ごとにそれぞれの質量と体積をもち，混合する気体の特性が異なることに対応し，気体定数は互いに異なっている．さらに，混合に際しては，仕事の付与や熱の移動がないものとする．

名称が i 番となる気体は，それ自身の状態方程式に従うことになり，混合する気体全体としては，式（10-1）に示した関係が存在している．

$$PV_i = G_i R_i T \left[\text{for } i\text{-th gas} \right], \quad G = \sum_{i=1}^{n} G_i, \quad V = \sum_{i=1}^{n} V_i, \quad \Delta W^G = 0, \quad \Delta Q = 0 \qquad (10\text{-}1)$$

After the process of mixing, the partial pressure of each gas can be noted by Eq.10-2 and this fact is called as "Dalton's law of partial pressure".

混合操作が終了した後，それぞれの気体の分圧は式（10-2）のように記述することができ，この内容はドルトンの分圧の法則とよばれている．

$$PV_i = p_i V \Rightarrow p_i = P \frac{V_i}{V}, P = \sum_{i=1}^{n} p_i \ (P : \text{total pressure 全圧}, \ p_i : \text{partial pressure 分圧}) \ (10\text{-}2)$$

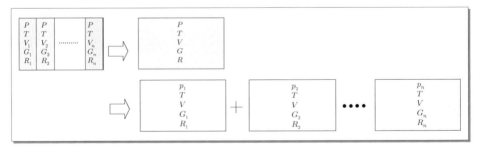

Fig.10-1 Mixing problem of gases under common pressure and temperature condition
圧力と温度が共通である条件下における気体の混合問題

According to the allocation principle for thermodynamic intensive quantities, the gas constant and specific heats of the mixture can be derived as follows:

熱力学的示強性量についての按分の考え方より，混合気体の気体定数や比熱については以下のように与えられている．

$$R_m = \frac{\sum_{i=1}^{n} G_i R_i}{G} \quad \Rightarrow \quad PV = GR_m T \tag{10-3}$$

$$c_{vm} = \frac{\sum_{i=1}^{n} G_i c_{vi}}{G}, \quad c_{pm} = \frac{\sum_{i=1}^{n} G_i c_{pi}}{G} \tag{10-4}$$

As for the case of Fig.10-2, when the pressures and temperatures of gases to be mixed are different, the temperature of the mixture can be derived using a balance of internal energy.

図 10-2 に示した場合では，混合すべき気体の圧力と温度がそれぞれ異なるので，混合気体の温度は，内部エネルギーのバランスから導出することができる．

$$\sum_{i=1}^{n} G_i c_{vi} T_i = T \sum_{i=1}^{n} G_i c_{vi} \tag{10-5}$$

$$T = \frac{\sum_{i=1}^{n} G_i c_{vi} T_i}{\sum_{i=1}^{n} G_i c_{vi}} = \frac{\sum_{i=1}^{n} \frac{1}{\kappa_i - 1} P_i V_i}{\sum_{i=1}^{n} \frac{1}{\kappa_i - 1} \frac{P_i V_i}{T_i}} \tag{10-6}$$

The partial pressure of each gas after mixing is given by Eq.10-7.

混合後の分圧は，式（10-7）により与えられている．

$$\frac{P_i V_i}{T_i} = \frac{p_i' V}{T} \quad \Rightarrow \quad p_i' = P \frac{V_i T}{V T_i},$$

$$P = \sum_{i=1}^{n} p_i' \ (P : \text{total pressure 全圧}, \ p_i : \text{partial pressure 分圧}) \tag{10-7}$$

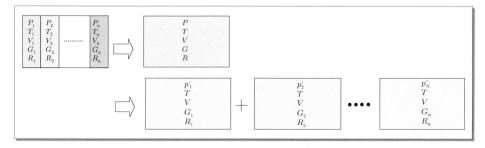

Fig.10-2　Mixing problem of gases with various pressures and temperatures
　　　　　種々の温度や圧力の気体の混合問題

10.2 Steady Flow Throttling

10.2 定常流における絞り

Steady flow throttling is a well-known problem in thermodynamics. Here, a gas of mass G [kg] flows with an upstream velocity of w_1 and passes through a throttle. Then, it flows downstream with a velocity of w_2. This flow system is illustrated in Fig.10-3.

定常流における絞りは，熱力学のよく知られている問題である．ここで，質量 G[kg] の気体が，上流から速度 w_1 で流れてきて絞りを通過する．その後，気体は速度 w_2 で下流に流れ去っていく．この流れの系は，図 10-3 に示したとおりである．

According to the energy balance of a flow, Eq.10-8 is always applicable for this flow system.

流れのエネルギーバランスによれば，この流れ系に対して式（10-8）がいつも成り立っている．

$$\dot{U}_1 + P_1\dot{V}_1 + \frac{1}{2}\dot{G}_1 w_1^2 + \dot{G}_1 g z_1 + \Delta\dot{Q} + \Delta\dot{U} = \dot{U}_2 + P_2\dot{V}_2 + \frac{1}{2}\dot{G}_2 w_2^2 + \dot{G}_2 g z_2 + \Delta\dot{W}^G \qquad (10\text{-}8)$$

From a comparison of the magnitudes of kinetic energy, potential energy and internal energy of the flow media, we can neglect the terms of kinetic energy and potential energy using Eq.10-9.

運動エネルギー，位置エネルギー，流体の内部エネルギーの三者の大きさのレベルの比較から，式（10-9）を用いて，運動エネルギーと位置エネルギーを無視することが可能である．

$$\frac{\frac{1}{2}\dot{G}w^2 + \dot{G}gz}{\dot{U}} \approx 0 \qquad (10\text{-}9)$$

Then, the energy equation for general flow can be reduced as follows:

したがって，一般的な流れについてのエネルギー式は，以下のように簡単化される．

$$\dot{U}_1 + P_1\dot{V}_1 + \Delta\dot{Q} + \Delta\dot{U} = \dot{U}_2 + P_2\dot{V}_2 + \Delta\dot{W}^G \qquad (10\text{-}10)$$

$$\dot{H}_1 + \Delta\dot{Q} + \Delta\dot{U} = \dot{H}_2 + \Delta\dot{W}^G \qquad (10\text{-}11)$$

Here, Eq.10-12 is introduced as a condition of steady and isolated flow.

さらに，定常で外部との熱や仕事の授受のない流れの条件として，式（10-12）を導入する．

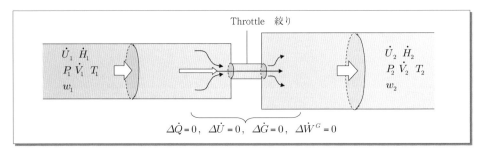

Fig.10-3 Steady flow throttling   定常流における絞り

$$\Delta \dot{Q} = 0, \quad \Delta \dot{U} = 0, \quad \Delta \dot{G} = 0, \quad \Delta \dot{W}^G = 0 \qquad (10\text{-}12)$$

Thus, iso-enthalpy flow can be derived for steady flow throttling.

その結果，定常流における絞りに対して，等エンタルピー流れであることが導出される．

$$\dot{H}_1 = \dot{H}_2 \qquad (10\text{-}13)$$

Equation 10-13 in the special case of enthalpy flow problem shown by Eq.5-44.

式（10-13）は，式（5-44）で示されたエンタルピー流れの特別な場合である．

10.3 Unsteady Throttling

10.3 非定常絞り

Consider a problem in which part of gas in a container is throttling out as shown in Fig.10-4. Throttling can be assumed as a process of quasi-equilibrium discharge and an adiabatic expansion of contained gas. Then, the fundamental equations of adiabatic expansion and mass conservation are satisfied during the process.

図 10-4 に示すような，容器内の気体の一部が絞りを介して流出する問題を検討する．絞りは準静的放出過程であり，かつ内部の気体の断熱膨張と想定することができる．そこで，断熱変化の基礎式と質量保存が絞り放出の過程において成立している．

$$dS = \frac{dQ}{T} = 0 \quad (\text{adiabatic expansion　断熱膨張}) \qquad (10\text{-}14)$$

$$G_1 = G_2 + G_d \qquad (10\text{-}15)$$

Since the pressure of discharged gas can be considered to be equivalent to the ambient pressure, Eq.10-17 is obtained for discharged gas.

放出された気体の圧力は，周囲の圧力と等しいと考えることができるので，以下の式を放出された気体について求めることができる．

$$P_d \equiv P_a = \text{const.} \quad (\text{ambient pressure　周囲圧力}) \qquad (10\text{-}16)$$

$$P_d V_d = P_a V_d = G_d P_a v_d = G_d R T_d \qquad (10\text{-}17)$$

Furthermore, Eq.10-18 and Eq.10-19 are derived for the initial gas and the gas left in the

さらに，最初の気体と容器内に残留した気体については，式（10-18）と式（10-19）が

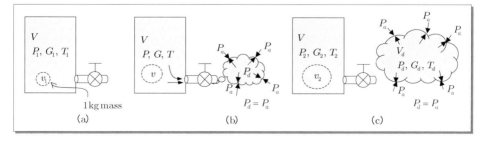

Fig. 10-4　Unsteady throttling　非定常絞り

container.   導かれる.

$$P_1 V = G_1 P_1 v_1 = G_1 R T_1, \quad P_2 V = G_2 P_2 v_2 = G_2 R T_2 \tag{10-18}$$

$$P_1 v_1^\kappa = P_2 v_2^\kappa, \quad \frac{T_1}{P_1^{\frac{\kappa-1}{\kappa}}} = \frac{T_2}{P_2^{\frac{\kappa-1}{\kappa}}} \tag{10-19}$$

From the energy balance before and after the discharge of gas, the following equation can be derived:

気体の放出の前後におけるエネルギーバランスから，つぎの式を導くことができる.

$$G_1 u_1 = G_2 u_2 + G_d u_d + \int_1^2 P_d dV = G_2 u_2 + G_d u_d + P_a V_d \tag{10-20}$$

Using the above equations, the following relationships can be derived for the gas discharged:

上記の式を用いることで，放出された気体についての以下の関係式を導くことができる.

$$P_a V_d = G_1 c_v T_1 - G_2 c_v T_2 - G_d c_v T_d \tag{10-21}$$

$$P_a V_d + G_d c_v T_d = G_1 c_v T_1 - G_2 c_v T_2 \tag{10-22}$$

$$\frac{\kappa}{\kappa-1} P_a V_d = \frac{1}{\kappa-1}(P_1 - P_2)V \tag{10-23}$$

$$P_a V_d = (P_1 - P_2)\frac{V}{\kappa} \tag{10-24}$$

As for the gas left in the container, Eq.10-25 can be derived.

容器内に残留している気体については，式（10-25）が導かれる.

$$P_2 V = (G_1 - G_d) R T_2 = \left(\frac{P_1 V}{T_1} - \frac{P_a V_d}{T_d} \right)\left(\frac{P_2}{P_1} \right)^{\frac{\kappa-1}{\kappa}} T_1 \tag{10-25}$$

Finally, we can derive the equation for the initial and final state of the gas in the container and the mass of discharged gas.

最終的に，容器内の最初と最後の状態と，放出された気体の質量との間の関係式が，以下のように得られる.

$$G_d = G_1 - G_2 = \left(\frac{P_1}{T_1} - \frac{P_2}{T_2} \right)\frac{V}{R} \tag{10-26}$$

10.4 Filling

10.4 充　填

Filling process is a kind of mixing process between gases inside and outside of a container. The typical filling process is shown in Fig.10-5. The total internal energy at the state before filling is given by Eq.10-27.

充填過程は，容器の内外の気体の混合過程の一つである．典型的な充填過程が，図 10-5 に示されている．充填前の全体の内部エネルギーは，式（10-27）で与えられている.

$$U_1 = G_1 c_v T_1 + G_f c_v T_f = c_v \frac{P_1 V}{R} + G_f c_v T_f \tag{10-27}$$

During the filling process, the gas that is filled receives compression work and its internal energy after filling increases and can be noted by Eqs.10-28 and 10-29.

充填の場合，充填される気体は圧縮仕事を受け取るので，充填後の内部エネルギーは増加し，式（10-28）と式（10-29）のように記述できる．

$$U_2 = U_1 + P_a V_f = U_1 + G_f R T_f \tag{10-28}$$

$$U_2 = (G_1 + G_f) c_v T_2 = c_v \frac{P_2 V}{R} \tag{10-29}$$

Uing Eq.10-27~Eq.10-29, the mass of the filling gas is given by Eqs.10-30 and 10-31.

式（10-27）〜（10-29）を用いると，充填される気体の質量は，式（10-30）と式（10-31）で与えられる．

$$G_f = \frac{(P_2 - P_1)V}{\kappa R T_f} \tag{10-30}$$

$$G_f = G_2 - G_1 = \frac{P_2 V}{R T_2} - \frac{P_1 V}{R T_1} \tag{10-31}$$

The temperature after filling can be obtained from Eq.10-30 and Eq.10-31.

充填後の温度は，式（10-30）と式（10-31）より得ることができる．

$$T_2 = \frac{\kappa T_1 T_f}{T_1 + (\kappa T_f - T_1)\dfrac{P_1}{P_2}} \tag{10-32}$$

When the initial temperatures of both the gases inside and outside of the container are equal, the temperature after filling is given by Eq.10-33.

容器の内外の気体について，両方の温度が等しい場合では，充填後の温度が式（10-33）で与えられる．

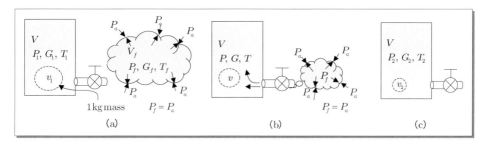

Fig.10-5　Filling　充填

$$T_f = T_1 \quad \Rightarrow \quad T_2 = \frac{\kappa T_1}{1 + (\kappa - 1)\dfrac{P_1}{P_2}} = \frac{\kappa T_f}{1 + (\kappa - 1)\dfrac{P_1}{P_2}} \tag{10-33}$$

Furthermore, when the gas is filled to an empty container, temperature is given by Eq.10-34. It is clear that the temperature increase is the result of the work done from the ambient surroundings to the gas.

さらに，空の容器にガスを充填する場合では，温度が式（10-34）で与えられる．周囲雰囲気から，気体に対してなされた仕事の結果として，温度が増加していることが明らかである．

$$P_1 = 0 \quad \Rightarrow \quad T_2 = \kappa T_f \tag{10-34}$$

10.5 Discharge and Filling

10.5 放出と充填

Consider a process in which a gas of mass G is transferred from container A to container B. This process is considered as a coupled process of discharge and filling as shown in Fig.10-6.

質量 G の気体を，容器 A から容器 B へ移動させる変化過程を検討する．この変化過程は，図 10-6 に示すように，放出と充填が対になっている変化過程とみなすことができる．

The mass of gas transferred from container A to container B is given by Eq.10-35.

容器 A から容器 B へ移動する気体の質量は，式（10-35）で与えられている．

$$G = G_{a1} - G_{a2} = G_{b2} - G_{b1} = (G_{bb2} + G_{ba2}) - G_{b1} = (G_{b1} + G_{ba2}) - G_{b1} = G_{ba2} \tag{10-35}$$

Assuming that the gases contained in the containers are the same kind but at different states, and considering that the process is adiabatic, the temperature, pressure and internal energy can then be given by the following equations:

同じ種類ではあるが，熱力学的状態の異なる気体が両方の容器に充填されているとし，かつ，変化過程は断熱過程であるので，温度，圧力，内部エネルギーは，以下の式で与えられる．

$$T_{b2} = T_{ba2} = T_{bb2}, \quad P_{b2} = P_{ba2} = P_{bb2}, \quad u_{b2} = u_{ba2} = u_{bb2} \tag{10-36}$$

$$G_{a1}u_{a1} + G_{b1}u_{b1} = G_{a2}u_{a2} + G_{b2}u_{b2} = (G_{a1} - G)u_{a2} + (G_{b1} + G)u_{b2} \tag{10-37}$$

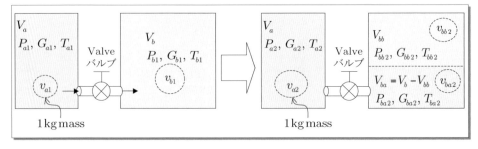

Fig.10-6 Discharge and filling   放出と充填

$$ds = \frac{dq}{T} = 0 \quad \left[\text{for gas "a" in } V_a\right] \quad \Rightarrow \quad \frac{T_{a1}}{P_{a1}^{\frac{\kappa-1}{\kappa}}} = \frac{T_{a2}}{P_{a2}^{\frac{\kappa-1}{\kappa}}} \tag{10-38}$$

Using Eq.10-37, the following equation can be derived:

式 (10-37) を用いれば，つぎの式を導出することができる．

$$P_{a1}V_a + P_{b1}V_b = P_{a2}V_a + P_{b2}V_b \tag{10-39}$$

Thus, as for the final state of the process, the following thermodynamic quantities are derived:

結果として，最終状態としての以下の熱力学的諸量が求められる．

$$P_{a2} = (G_{a1} - G)\frac{RT_{a2}}{V_a} = (G_{a1} - G)\frac{RT_{a1}}{V_a}\left(\frac{P_{a2}}{P_{a1}}\right)^{\frac{\kappa-1}{\kappa}} \tag{10-40}$$

$$P_{a2}V_a = P_{a1}V_a\left[1 - \frac{T_{a1}V_b}{P_{a1}V_a}\left(\frac{P_{b2}}{T_{b2}} - \frac{P_{b1}}{T_{b1}}\right)\right]^{\kappa} \tag{10-41}$$

$$G = \frac{V_b}{R}\left(\frac{P_{b2}}{T_{b2}} - \frac{P_{b1}}{T_{b1}}\right) = \frac{V_a}{R}\left(\frac{P_{a1}}{T_{a1}} - \frac{P_{a2}}{T_{a2}}\right) \tag{10-42}$$

Here, when the container B is initially empty, the final pressures and temperatures of both containers should satisfy the following equations.

ここで，最初に容器 B が空であるならば，両容器の最終的な圧力と温度は，以下の式を満足する．

$$V_a = V_b, \quad P_b = 0 \quad \Rightarrow \quad P_{a1} \geq P_{a2} \geq P_{b2}, \quad T_{a2} \leq T_{a1} \leq T_{b2} \tag{10-43}$$

Problems

△10-1

Consider a gas mixture that consists of 21 vol% O_2 and 79 vol%N_2. If the molar weights of O_2 and N_2 are 32.00 kg/kmol and 28.02 kg/kmol respectively, determine (1) the average molar weights of the mixture, (2) the mass fraction of each component. Moreover, determine (3) the average gas constant of the mixture, if the gas constants of O_2 and N_2 are 0.2598 kJ/(kg·K) and 0.2968 kJ/(kg·K), respectively.

△10-2

If P_1 = 1 MPa, V = 10 m³, T_1 = 300 K, P_2 = 0.5 MPa, and P_a = 0.1 MPa in Fig.10-7, determine the volume and mass of discharged gas V_d and G_d. The gas constant and ratio of specific heat are 0.287 kJ/(kg·K) and 1.40 respectively.

問題

△10-1

21 vol% の O_2 と 79 vol% の N_2 の混合ガスを考える．O_2 と N_2 の分子量をそれぞれ 32.00 kg/kmol, 28.02 kg/kmol とするとき，(1) 混合ガスの平均分子量，(2) 各成分の質量分率をそれぞれ求めなさい．さらに，(3) O_2 と N_2 のガス定数をそれぞれ 0.2598 kJ/(kg·K), 0.2968 kJ/(kg·K) とするときの混合ガスのガス定数を求めなさい．

△10-2

図 10-7 において，P_1 = 1 MPa, V = 10 m³, T_1 = 300 K, P_2 = 0.5 MPa, P_a = 0.1 MPa であるとき，放出されたガスの体積 V_d と質量 G_d を求めなさい．ガス定数と比熱比は，それぞれ 0.287 kJ/(kg·K) と 1.40 とする．

△ 10-3

If P_1 = 0.2 MPa, V = 3 m³, T_1 = 300 K, T_f = 280 K and P_2 = 10 MPa in Fig.10-8, determine T_2 and G_f. The gas constant and the ratio of specific heat are 0.287 kJ/(kg·K) and 1.40, respectively.

△ 10-3

図 10-8 において，P_1 = 0.2 MPa, V = 3 m³, T_1 = 300 K, T_f = 280 K, P_2 = 10 MPa であるとき，T_2 と G_f を求めなさい．ガス定数と比熱比はそれぞれ 0.287 kJ/(kg·K) と 1.40 とする．

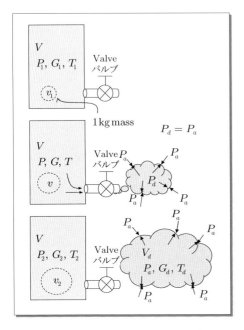

Fig.10-7   Schematic for Problem 10-2
　　　　　 問題 10-2 の説明図

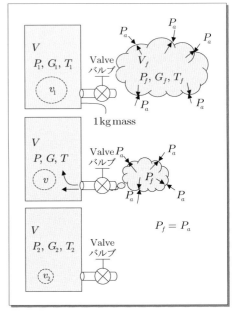

Fig.10-8   Schematic for Problem 10-3
　　　　　 問題 10-3 の説明図

Chapter 11
Nozzle Flow
ノズル流れ

11.1 Fundamental Equations of Flow

11.1 流れの基礎方程式

A general scheme of a flow in a duct is illustrated in Fig.11-1. It has an entrance and an exit of flow. Here, we assume that thermal energy is supplied externally and that external work is done by the flow media in the duct. Furthermore, there are internal sources of heat and mass inside the duct. A mass balance including an internal mass source is given by Eq.11-1.

図 11-1 に，ダクト内の流れの一般的な模式図を示す．ダクトには，流れの入口と出口がある．熱としてのエネルギーは外部から供給されるものとし，また，仕事も流れによって行われるものとする．ダクト内では，さらにその内部で新たな発熱と質量の湧き出しが行われているものとする．内部での質量の湧き出しを含む質量バランスは，式（11-1）で与えられている．

$$\dot{G}_1 + \Delta\dot{G} = \dot{G}_2, \quad A_1 \frac{1}{v_1} w_1 + \Delta\dot{G} = A_2 \frac{1}{v_2} w_2 \tag{11-1}$$

This equation is commonly known as the equation of continuity. A differential form of this equation shown by Eq.11-2 is the general equation of continuity.

この式は，連続の式としてよく知られているものである．式（11-2）に示したこの式の微分形は，さらに一般的な連続の式である．

$$d\left(A \cdot \frac{1}{v} \cdot w\right) + d\dot{G} = 0$$

$$\Rightarrow \quad \frac{1}{v} w dA - Aw \frac{dv}{v^2} + A \frac{1}{v} dw + d\dot{G} = 0 \quad \Rightarrow \quad \frac{dA}{A} - \frac{dv}{v} + \frac{dw}{w} + \frac{v d\dot{G}}{Aw} = 0 \tag{11-2}$$

Assuming $\Delta\dot{G} = 0$ and $\Delta\dot{Q}_S = 0$ in the duct, for simplification, an energy balance including all terms of heat and work is given by Eq.11-3.

簡単にするため，ダクト内で$\Delta\dot{G} = 0$と$\Delta\dot{Q}_S = 0$とすれば，熱と仕事のすべての項を含むエネルギーバランスは，式（11-3）で与

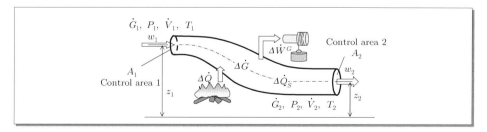

Fig.11-1 Concept of duct flow with external heating and work
外部加熱や外部仕事をともなうダクト流れの概念

えられている.

$$\dot{U}_1 + P_1\dot{V}_1 + \frac{1}{2}\dot{G}_1w_1^2 + \dot{G}_1gz_1 + \varDelta\dot{Q} = \dot{U}_2 + P_2\dot{V}_2 + \frac{1}{2}\dot{G}_2w_2^2 + \dot{G}_2gz_2 + \varDelta\dot{W}^G \qquad (11\text{-}3)$$

The internal energy change which occurs between the entrance and exit is given by Eq.11-4, and its differential form becomes Eq.11-5.

入口と出口の間で生じる内部エネルギーの変化は, 式 (11-4) で与えられる. さらに, その微分形式は式 (11-5) のようになる.

$$\left(\dot{U}_2 - \dot{U}_1\right) + \left(P_2\dot{V}_2 - P_1\dot{V}_1\right) + \frac{1}{2}\left(\dot{G}_2w_2^2 - \dot{G}_1w_1^2\right) + g\left(\dot{G}_2z_2 - \dot{G}_1g\right) + \varDelta\dot{Q} - \varDelta\dot{W} \qquad (11\text{-}4)$$

$$d\dot{U} + d\left(P\dot{V}\right) + \frac{1}{2}d\left(\dot{G}w^2\right) + gd\left(\dot{G}z\right) + d\dot{Q} - d\dot{W}^G = 0 \qquad (11\text{-}5)$$

As for a flow of unit mass of gas, the equations of thermodynamic quantities can be simply expressed as follows:

単位質量の気体の流れについて, 熱力学的諸量の関係方程式は, 以下のように簡単化される.

$$du + d(Pv) + wdw + gdz + dq - dW = 0 \qquad (11\text{-}6)$$

$$dh + wdw + gdz + dq - dW = 0 \qquad (11\text{-}7)$$

When the flow is steady and no heat and work are related to the flow, Eq.11-8 is suitable for the flow.

流れが定常で, 熱や仕事と流れが無関係の状態であるならば, 式 (11-8) が流れに対して適用される.

$$dz = 0, \quad dq = 0, \quad dW = 0 \;\Rightarrow\; dh + wdw = 0 \qquad (11\text{-}8)$$

Using the definition of enthalpy, Eq.11-9 is obtained for the condition where no heat is supplied.

エンタルピーの定義に従えば, 熱の供給がない場合では式 (11-9) が得られる.

$$dq = 0 \;\Rightarrow\; dh - vdP = 0 \qquad (11\text{-}9)$$

Finally, the equation of duct flow becomes Eq.11-10.

最終的に, ダクト内の流れの方程式は式 (11-10) となる.

$$vdP + wdw = 0 \qquad (11\text{-}10)$$

11.2 Sound Velocity

11.2 音　速

Sound is a perturbation pressure wave propagating in a substance. Then, consider the sound velocity in a gaseous media using a model (Fig.11-2) of plane wave and adiabatic pressure change.

The equation of continuity under the condition of a plane wave model with no mass change can be noted simply as indicated by Eq.11-11.

音は物質の中を伝わっていく微小変動の圧力波である. そこで, 平面波のモデル (図11-2) と断熱圧力変化を用いて, 気体状の媒体中の音速を検討する.

質量の増減がないとする平面波の条件下において, 連続の式は式 (11-11) で示されるように簡単化される.

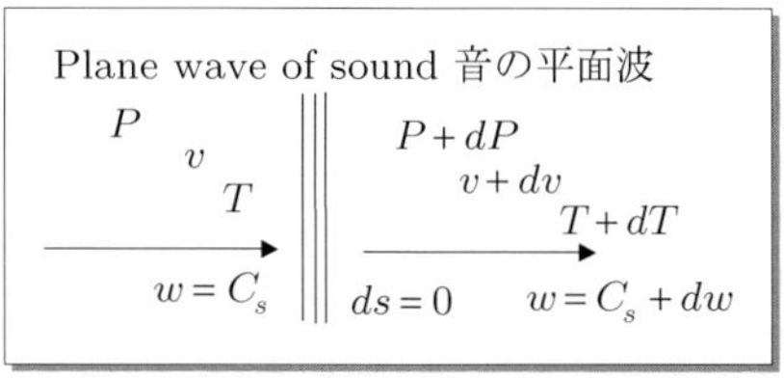

Fig.11-2　Pressure wave and sound　圧力波と音

$$\frac{dA}{A} - \frac{dv}{v} + \frac{dw}{w} + \frac{vdG}{Aw} = 0, \quad dA = 0, \quad dG = 0 \ \Rightarrow\ \frac{dw}{w} - \frac{dv}{v} = 0 \tag{11-11}$$

As for an energy balance, no change of internal energy is assumed and Eq.11-13 is obtained from a modification of the fundamental equation shown in Eq.11-12.

エネルギーバランスについては，内部エネルギーが変化しないことを想定すると，式（11-12）に示した基礎式の修正式から，式（11-13）が得られる．

$$du + d(Pv) + wdw + gdz + dq - dW = 0, \quad du = 0, \quad dz = 0, \quad dq = 0, \quad dW = 0$$
$$\Rightarrow\ d(Pv) + wdw = 0 \tag{11-12}$$

$$\frac{d(Pv)}{w^2} + \frac{dv}{v} = 0 \tag{11-13}$$

The equation of the wave velocity is then expressed by Eq.11-14.

波の速度は，結果として式（11-14）で表されることになる．

$$C_s^2 = w^2 = -\frac{v}{dv}d(Pv) \tag{11-14}$$

Since the sound wave is a perturbation wave of pressure, and is adiabatic with no internal energy change, we can use the following assumption:

音の波は断熱かつ内部エネルギーの変化のない圧力の微小擾乱波であることから，つぎのような仮定を使うことが可能である．

$$\text{for sound velocity} \quad dw = 0 \ \Rightarrow\ Pdv = 0 \ \Rightarrow\ d(Pv) = vdP \tag{11-15}$$

Furthermore, using the equation of state and condition of adiabatic change, we can derive the sound velocity.

さらに，状態方程式と断熱変化の条件を用いることで，音速を求めることができる．

$$Pv = RT, \quad ds = 0 \ \Rightarrow\ c_p\frac{dv}{v} + c_v\frac{dP}{P} = 0 \tag{11-16}$$

$$C^2 = -v^2\frac{dP}{dv} = \frac{c_p}{c_v}Pv \tag{11-17}$$

$$C = \sqrt{\kappa Pv} = \sqrt{\kappa RT} \quad (\text{sound velocity　音速}) \tag{11-18}$$

11.3 Nozzle Flow

Consider a gaseous flow issuing from a nozzle as shown in Fig.11-3. We start our analysis with the fundamental energy balance assuming adiabatic flow conditions. It means that Eq.11-19 is the fundamental equation of the analysis.

With adiabatic conditions considered, the velocity change between the upstream region and the nozzle exit is expressed by the enthalpy change between them. Also the adiabatic change shows the relationship between the temperature and pressure as shown in Eq.11-20.

Then, the velocity change can be expressed as follows:

As for the upstream region, the upstream velocity is low enough to be negligible. Thus, the flow velocity at the nozzle exit can be noted as follows:

11.3 ノズル流れ

図 11-13 に示すようなノズルから流出する流れを検討する．基本的なエネルギーバランスと断熱流れの条件を解析の前提とする．このことは，式（11-19）が解析の基礎方程式であることと同じである．

$$du + d(Pv) + \frac{1}{2} w^2 + gdz + dq + dW = 0, \quad dz = 0, \quad dq = 0, \quad dW = 0 \qquad (11\text{-}19)$$

断熱であるという条件により，上流の流れ場とノズルの出口との間の速度の変化は，両者の間のエンタルピー変化で表現することができる．さらに，断熱変化であることは，式（11-20）に示した温度と圧力の関係があることを示している．

$$\frac{w^2 - w_1^2}{2} = h_1 - h \qquad h_1 - h = c_p (T_1 - T) \qquad \frac{T}{T_1} = \left(\frac{P}{P_1} \right)^{\frac{\kappa - 1}{\kappa}} \qquad (11\text{-}20)$$

したがって，速度変化はつぎの式で表現される．

$$w^2 - w_1^2 = 2(h_1 - h) = 2c_p T_1 \left\{ 1 - \left(\frac{P}{P_1} \right)^{\frac{\kappa - 1}{\kappa}} \right\} = 2 \frac{\kappa}{\kappa - 1} P_1 v_1 \left\{ 1 - \left(\frac{P}{P_1} \right)^{\frac{\kappa - 1}{\kappa}} \right\} \qquad (11\text{-}21)$$

ノズルの上流に関しては，上流の速度は無視できるほど小さいので，結果としてノズル出口の速度はつぎのようになる．

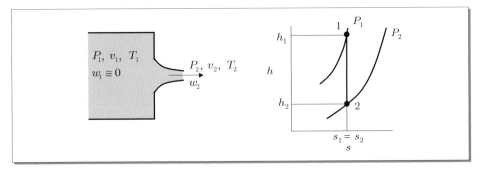

Fig.11-3 Adiabatic nozzle   断熱ノズル

$$w_1 \cong 0 \tag{11-22}$$

$$w_2 = \sqrt{2\frac{\kappa}{\kappa-1}P_1 v_1 \left\{1-\left(\frac{P_2}{P_1}\right)^{\frac{\kappa-1}{\kappa}}\right\}} \tag{11-23}$$

11.4 Nozzle Flow with Energy Loss

Since there are some energy losses such as nozzle wall friction, the actual velocity of the nozzle exit is less than the theoretical velocity obtained by Eq.11-23. On the h-s chart, it is expressed by entropy and enthalpy increases during the discharge phenomena as indicated by Fig.11-4.

We can start the analysis of this problem using Eq.11-24 which is exactly the same as Eq.11-19, but $dW \neq 0$ is assumed.

11.4 エネルギーの損失をともなう ノズル流れ

ノズル壁面での摩擦のようなエネルギー損失があるので，実際のノズルからの流出速度は，式（11-23）で得られる理論的な値より低くなる．h-s 線図上において，これは流出の間のエンタルピーとエントロピーの増加として，図 11-4 のように表されている．

この問題は $dW \neq 0$ ではあるが，式（11-19）とまったく同じである式（11-24）を基礎として解析を行うことができる．

$$du+d(Pv)+\frac{1}{2}w^2+gdz+dq+dW=0, \quad dz=0, \quad dq=0, \quad dW \neq 0 \tag{11-24}$$

Using the nozzle efficiency η, the enthalpy change during the discharge can be noted as follows. Here, the polytropic change is assumed instead of adiabatic change.

ノズル効率 η を用いることで，流出にともなうエンタルピー変化はつぎのように表すことができる．なおここでは，断熱変化の代わりにポリトロープ変化が想定されている．

$$w_1 \cong 0 \qquad h'-h_1 = \phi^2(h'-h_1) \qquad w'=\phi w \qquad Pv^n = \text{const.} \tag{11-25}$$

$$w_2' = \sqrt{2\frac{\kappa}{\kappa-1}P_1 v_1 \left\{1-\left(\frac{P_2}{P_1}\right)^{\frac{n-1}{n}}\right\}} = \phi \sqrt{2\frac{\kappa}{\kappa-1}P_1 v_1 \left\{1-\left(\frac{P_2}{P_1}\right)^{\frac{\kappa-1}{\kappa}}\right\}} \tag{11-26}$$

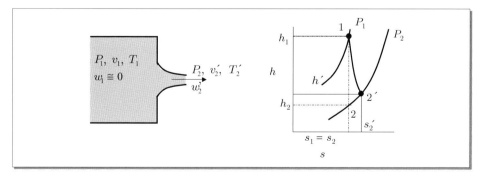

Fig.11-4　Nozzle with energy loss　エネルギー損失をともなうノズル

$$\eta_n = \phi^2 = \frac{h_1 - h_2'}{h_1 - h_2} = \frac{w_2'^2}{w_2^2} = \left\{ 1 - \left(\frac{P_2}{P_1} \right)^{\frac{n-1}{n}} \right\} \Big/ \left\{ 1 - \left(\frac{P_2}{P_1} \right)^{\frac{\kappa-1}{\kappa}} \right\} \tag{11-27}$$

11.5 Critical Pressure for Nozzle Flow

Mass flow issuing from the nozzle is derived by Eq.11-28 when the cross-sectional area of the nozzle is noted by A_2.

11.5 ノズル流れとその臨界圧力

ノズルの流出断面積が A_2 で与えられているならば，ノズルから流出する質量は，式（11-28）で与えられる．

$$\dot{G} = A_2 \frac{1}{v_2} w_2 = A_2 \left(\frac{P_2}{P_1} \right)^{\frac{1}{\kappa}} \frac{w_2}{v_1} = A_2 \left(\frac{P_2}{P_1} \right)^{\frac{1}{\kappa}} \sqrt{2 \frac{\kappa}{\kappa-1} \frac{P_1}{v_1} \left\{ 1 - \left(\frac{P_2}{P_1} \right)^{\frac{\kappa-1}{\kappa}} \right\}}$$

$$= A_2 \sqrt{2 \frac{\kappa}{\kappa-1} \frac{P_1}{v_1} \left\{ \left(\frac{P_2}{P_1} \right)^{\frac{2}{\kappa}} - \left(\frac{P_2}{P_1} \right)^{\frac{\kappa+1}{\kappa}} \right\}} \tag{11-28}$$

It is clear that the mass flow depends on the pressure ratio between the upstream pressure and downstream pressure of the nozzle. The maximum mass flow rate is given by the analysis shown in Eq.11-29.

質量流量がノズルの上流と下流の圧力比によって与えられていることが，この式より明らかである．さらに，流出質量の最大値は式（11-29）に示した解析により与えられる．

$$\frac{d\dot{G}}{d\left(\frac{P_2}{P_1} \right)} = 0 \;\Rightarrow\; \frac{P_2}{P_1} = \left(\frac{2}{\kappa+1} \right)^{\frac{\kappa}{\kappa-1}} = \frac{P_c}{P_1} \text{ (critical pressure for nozzle flow ノズル流れの臨界圧力)} \tag{11-29}$$

Eq.11-29 gives the critical flow of the nozzle. When the pressure ratio between the upstream and downstream sides of the nozzle is higher than the critical value, the mass flow rate is given by Eq.11-30. However, since the flow velocity does not increase beyond the sound velocity, the mass flow rate is fixed on the critical value indicated by Eq.11-31 when the ratio is less than this critical value.

式（11-29）は，ノズルの臨界流れを与えている．ノズルの上流と下流の圧力比が臨界値よりも大きい場合では，質量流量は式（11-30）で与えられる．しかし，流れの速度が音速を超えることはないので，この圧力比が限界値より小さな場合では，質量流量は式（11-31）で示されている臨界値に固定されてしまう．

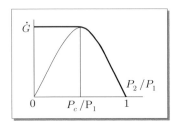

Fig.11-5 Issuing mass from nozzle
ノズルからの流出質量

$$\frac{p_c}{P_1} < \frac{p_2}{P_1} < 1 \qquad \dot{G} = A_2 \sqrt{2 \frac{\kappa}{\kappa-1} \frac{P_1}{v_1} \left\{ \left(\frac{P_2}{P_1}\right)^{\frac{2}{\kappa}} - \left(\frac{P_2}{P_1}\right)^{\frac{\kappa+1}{\kappa}} \right\}} \qquad (11\text{-}30)$$

$$0 < \frac{p_2}{P_1} \leq \frac{p_c}{P_1} \qquad \dot{G} = A_2 \sqrt{2 \frac{\kappa}{\kappa+1} \left(\frac{2}{\kappa+1}\right)^{\frac{2}{\kappa-1}} \frac{P_1}{v_1}} \qquad (11\text{-}31)$$

11.6 Supersonic Flow

From Eq.11-23 and Eq.11-29, the issuing velocity at the critical pressure condition is given by Eq.11-32.

11.6 超音速流

式 （11-23） と式 （11-29） より，臨界条件下では流出速度が式 （11-32） で与えられる.

$$w_c = \sqrt{2 \frac{\kappa}{\kappa-1} P_1 v_1 \left\{ 1 - \left(\frac{2}{\kappa+1}\right) \right\}} = \sqrt{2 \frac{\kappa}{\kappa+1} \left(\frac{P_1}{P_c}\right)^{\frac{\kappa-1}{\kappa}} P_c v_c} = \sqrt{\kappa P_c v_c} = C_c = C_s \qquad (11\text{-}32)$$

It suggests that the critical velocity is absolutely coincident to the sound velocity at the nozzle exit.

Here, three kinds of nozzle configurations, which are basically considered as typical nozzles, are illustrated in Fig.11-6. The maximum issuing velocity of a convergent nozzle is the sound velocity because no perturbation pressure wave can propagate upstream with a velocity higher than the sound velocity even if the pressure ratio is less than the critical value.

Divergent nozzles and Laval nozzles are used to accelerate the flow above the sound velocity. Laval nozzle is one of the more effective nozzles to realize a supersonic flow. For a supersonic flow, the nozzle configuration and flow conditions are expressed by the following equations. The outline of the pressure and velocity changes in the nozzle is shown in Fig.11-7.

この式は臨界速度がノズル出口における音速そのものであることを示している.

ここで，図 11-6 に図示されている 3 種類の形状のノズルを基本的なノズルとして考えることができる. 先細ノズルの最大噴出速度は音速である. このことは，圧力比が臨界値よりも低い場合であっても音速より速い速度で圧力変動が上流に伝わることがないということに対応している.

末広ノズルとラバールノズルは，音速以上に流れを加速する場合に使われている. ラバールノズルは，超音速流を実現するための効果的なノズルの一つである. 超音速の流れに関連するノズルの形状と流れの条件は以下の式で示されている. また，ノズル内の圧力や速度の変化の状態は，図 11-7 に示したようになる.

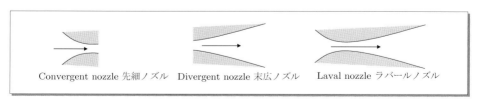

Convergent nozzle 先細ノズル　Divergent nozzle 末広ノズル　Laval nozzle ラバールノズル

Fig.11-6　Three kinds of basic nozzles　3 種類の基本ノズル

$$\frac{Aw}{v} = \text{const.}, \quad Pv^{\kappa} = \text{const.}, \quad wdw = -vdP \tag{11-33}$$

$$\frac{dA}{A} = \frac{dv}{v} - \frac{dw}{w}, \quad \frac{dv}{v} = -\frac{dP}{\kappa P}, \quad \frac{dw}{w} = -\frac{vdP}{w^2} \tag{11-34}$$

$$\frac{dA}{A} = -\frac{dP}{\kappa P} + \frac{vdP}{w^2} = -\frac{dP}{\kappa P}\left(1 - \frac{\kappa Pv}{w^2}\right)$$

$$= -\frac{dP}{\kappa P}\left(1 - \frac{C_c^2}{w^2}\right) = -\frac{dP}{\kappa P}\left(1 - \frac{1}{M^2}\right) \quad (M:\text{ Mach number } \ \text{マッハ数}) \tag{11-35}$$

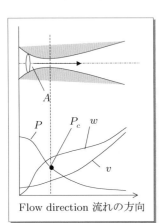

Flow direction 流れの方向

Fig.11-7 Laval nozzle and supersonic flow
ラバールノズルと超音速流

Problems	問題

△11-1

Air at 20 °C flows with a velocity of 300 m/s. If the gas constant and ratio of specific heat are 0.287 kJ/(kg·K) and 1.40, respectively, determine the velocity of sound and Mach number for the air flow.

△11-2

If $P_1 = 0.2$ MPa, $T_1 = 300$ K, and $P_2 = 0.1$ MPa in Fig.11-8, determine w_2. The gas constant and ratio of specific heat are 0.287 kJ/(kg·K) and 1.40, respectively.

△11-3

The actual value of w_2 is 320 m/s in the case of Problem 11-2. Determine the nozzle efficiency in this case.

△11-4

$P_1 = 0.5$ MPa, $T_1 = 30$ °C, and $A_2 = 400$ mm^2 in

△11-1

20 °C の空気が 300 m/s で流れている．ガス定数と比熱比をそれぞれ 0.287 kJ/(kg·K)，1.40 とするとき，空気中の音速と空気流れのマッハ数を求めなさい．

△11-2

図 11-8 において，$P_1 = 0.2$ MPa, $T_1 = 300$ K, $P_2 = 0.1$ MPa であるとき，w_2 を求めなさい．ガス定数と比熱比は，それぞれ 0.287 kJ/(kg·K) と 1.40 とする．

△11-3

問 11-2 の場合，実際の w_2 の値は 320 m/s であった．このときのノズル効率を求めなさい．

△11-4

図 11-8 において，$P_1 = 0.5$ MPa, $T_1 = 30$ °C, A_2

Fig.11-8. Determine the mass flow rate through the nozzle (1) at $P_2 = 0.3$ MPa and (2) at $P_2 = 0.1$ MPa. The gas constant and ratio of specific heat are 0.287 kJ/(kg·K) and 1.40, respectively.

= 400 mm² である．(1) $P_2 = 0.3$ MPa の場合のノズルを通る質量流量，(2) $P_2 = 0.1$ MPa の場合のノズルを通る質量流量を求めなさい．ガス定数と比熱比はそれぞれ 0.287 kJ/(kg·K) と 1.40 とする．

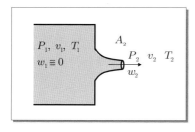

Fig.11-8 Schematic for Problems
問題の説明図

Chapter **12**
Thermodynamic Cycle and Entropy
熱力学的サイクルとエントロピー

12.1 Thermodynamic Cycle

Using an illustration shown by Fig.12-1, consider a thermodynamic process that starts from state 1 and reaches state 2 passing through a process "a", and returns to state 1 passing through a process "b". When process "b" is absolutely consistent with process "a", process "b" means the reverse process of "a". On the other hand, when process "b" is different with "a", the total process is called a cycle. The supplied quantities of heat and the works done during the cycle are illustrated in Fig.12-1.

Here, we assume an ideal gas of 1 kg mass as the working gaseous media in the cycle.

$$Pv = RT \tag{12-1}$$

The energy balance among supplied heat, internal energy and work shown in Eq.12-2 can be applied throughout the whole process.

$$dq = du + Pdv \tag{12-2}$$

Then, Eq.12-3 and Eq.12-4 can be derived for the processes of "a" and "b", respectively.

12.1 熱力学的サイクル

図 12-1 に示した説明図を用いて，状態 1 から出発し，"a" の過程を経て状態 2 に達したのち，"b" の過程を経たあとに状態 1 にもどる熱力学的な変化を検討する．ここでもし，"b" の過程が "a" の過程と完全に一致しているのならば，"b" の過程は "a" の過程の逆方向過程である．一方，"b" の過程が "a" の過程と異なる場合では，全体の変化過程がサイクルとよばれている．サイクルの間に供給される熱量とサイクルのする仕事については，図 12-1 に示してある．

ここで，質量 1 kg の理想気体を作動気体として想定する．

供給された熱と内部エネルギーと仕事の間のエネルギーバランスを式（12-2）に示しているが，この式は，変化過程の全域で成立する．

したがって，過程 "a" と過程 "b" について，式（12-3）と式（12-4）がそれぞれ成り立つ．

$$q_{1-a-2} = q_a = \int_{1-a}^{2} du + \int_{1-a}^{2} Pdv = u_2 - u_1 + \int_{1-a}^{2} Pdv \tag{12-3}$$

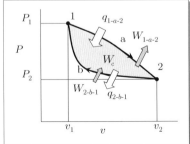

Fig.12-1 Thermodynamic cycle
熱力学的サイクル

$$q_{2-b-1} = q_b = \int_{2-b}^{1} du + \int_{2-b}^{1} Pdv = u_1 - u_2 + \int_{2-b}^{1} Pdv \qquad (12\text{-}4)$$

Here, a positive value of heat represents supplied heat and a negative value indicates released heat. As for work, positive work means work done by the cycle and negative work represents received work. Since both internal energies for the initial state and the final state are the same and are noted by "u_1", the total quantity of heat supplied and released is given by Eq.12-5.

ここで，正の値の熱はサイクルに供給された熱量であり，負の値は放熱である．仕事についていえば，正の仕事はサイクルが行った仕事を，そして負の値はその逆の仕事を意味している．サイクルの最初の状態と，最後の状態の内部エネルギーは同じであり，u_1 で表されているので，サイクルに供給，またサイクルから放出された全熱量は，式（12-5）で与えられる．

$$q_{1-a-2-b-1} = q_{1-a-2} + q_{2-b-1} = \int_{1-a}^{2} Pdv + \int_{2-b}^{1} Pdv = \oint_{1-a-2-b-1} Pdv \qquad (12\text{-}5)$$

$$q_c = q_{1-a-2-b-1} \qquad (12\text{-}6)$$

As for work, the total work can be noted by Eq.12-7. It is clear on the P-v chart that the total work corresponds to the area enclosed by the cycle. It also represents the substantial work produced by the cycle.

仕事に関しては，全仕事を式（12-7）のように表すことができる．この全仕事がサイクルで囲まれた面積に等しいことが P-v 線図上で明らかである．これはまた，サイクルで生じる実質的な仕事を表している．

$$W_{1-a-2-b-1} = W_{1-a-2} + W_{2-b-1} = \oint_{1-a-2-b-1} Pdv = \big[\,\text{area 1-a-2-b-1}\,\big] \qquad (12\text{-}7)$$

$$W_c = W_{1-a-2-b-1} \qquad (12\text{-}8)$$

Consequently, the energy balance can be noted by Eq.12-9.

結果として，エネルギーバランスは，式（12-9）で与えられている．

$$q_c = W_c \qquad (12\text{-}9)$$

The concept of thermal efficiency is introduced here to show the performance of the cycle. Thermal efficiency is defined as the conversion ratio of the substantial work produced by the cycle to the quantity of heat received. It is defined by Eq.12-10.

ここで，サイクルの性能を示すために熱効率の概念を取り入れる．熱効率の定義は，サイクルが受け取った熱量から実質的な仕事への変換率であり，これは式（12-10）で定義されている．

$$\eta = \frac{W_c}{q_a} = \frac{q_c}{q_a} = \frac{|q_a| - |q_b|}{|q_a|} = 1 - \left|\frac{q_b}{q_a}\right| \qquad (12\text{-}10)$$

Various cycles are shown in Fig.12-2. In case (a), states 1 and 2 are considered to be on the same adiabatic line and position "a" is in the upper position of the adiabatic line. Process 1-a-2 means an expansion process with positive work.

図 12-2 に，種々のスタイルのサイクルを示す．図（a）の場合，状態 1 と状態 2 は同じ断熱変化線上にあるとする．さらに "a" の位置は断熱線より上側にあるとする．1-a-2 の変化過程は正の仕事をともなう膨張過程で

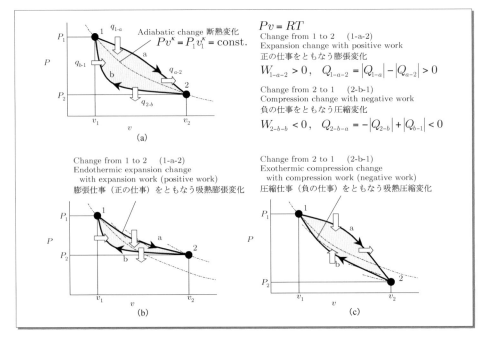

Fig.12-2   Thermodynamic cycle with endothermic and exothermic changes
吸熱および放熱変化をともなう熱力学的サイクル

However, the former half of the change is endo-thermic and the later half is exothermic. As for the compression process shown by 2-b-1, the former half of the process is exothermic but the later half is endothermic. Case (b) is a cycle in which the expansion process consists of a unique endothermic change, and case (c) is the cycle with an endothermic compression process.

We have to notice that the quantity of heat q_a generally expressed by Eq.12-3 is the total of the endothermic and exothermic heat movements during the expansion process, and q_b is the total of the exothermic and endothermic heat move-ment during the compression process.

12.2 Cycle and Entropy Change

A detailed analysis of entropy change during a cycle can be explained using the cycle samples shown in Fig.12-2. P-v charts and T-s charts with heat movements and works for the three

あるが，その前半は吸熱過程であり，その後半は放熱過程となる．2-b-1 で示されている圧縮変化過程では，変化過程の前半は放熱過程であり，後半は吸熱過程となる．図（b）の場合は，膨張変化過程が吸熱過程のみからなるサイクルであり，図（c）の場合は圧縮変化過程が吸熱過程となるサイクルである．

一般的に，式（12-3）で表現されている q_a は，膨張変化過程におけるサイクルの吸熱と放熱を合わせた熱の移動であり，q_b は，圧縮変化過程における放熱と吸熱を合わせた熱の移動であることを理解しておく必要がある．

12.2 サイクルとエントロピー変化

サイクルにおけるエントロピー変化の詳細は，図 12-2 に示したサイクルの例を使って説明することができる．図 12-2 の三つの例についての熱の動きや，仕事についての P-v

cases illustrated in Fig.12-2 are shown in Fig,12-3, Fig.12-4 and Fig.12-5, respectively.

As for the case of Fig.12-2 (a), gas states indicated by "1" and "2" are on the same adiabatic line and this means that there is no entropy change between states 1 and 2. The $T\text{-}s$ chart in Fig.12-3 shows that endothermic and exothermic changes can be clearly separated by the point "a" whose position indicates the maximum entropy point during an expansion process. Point "b" is also the separating point of exothermic and endothermic changes during the compression process. From the definition of entropy, the area in the $T\text{-}s$ chart enclosed by the cycle is the total quantity of heat consumed by the cycle and it is equivalent to the substantial work produced by the cycle.

線図と $T\text{-}s$ 線図を，三つの場合のそれぞれについて図 12-3，12-4，12-5 に示す．

図 12-2（a）の場合では，"1" と "2" で示されている気体の状態は，同じ断熱線上にある．このことは，状態 1 と状態 2 の間でエントロピーの変化がないことを示している．図 12-3 に示した $T\text{-}s$ 線図では，"a" 点により吸熱変化と放熱変化がはっきりと区別されている．ここで，"a" 点は膨張変化過程におけるエントロピーの最大点を示している．"b" 点もまた，圧縮変化過程における放熱過程と吸熱過程を区分する点になっている．エントロピーの定義から，サイクルで囲まれた $T\text{-}s$ 線図の面積はサイクルが消費した全熱量であり，これはサイクルにより生じた実質的な仕事と等しくなっている．

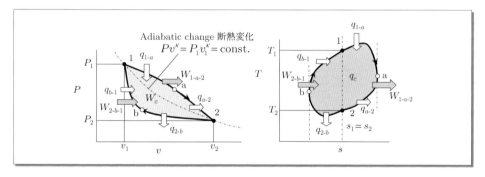

Fig.12-3　Cycle and entropy change for the case of Fig.12-2 (a)
　　　　　図 12-2(a) に対応したサイクルとエントロピー変化

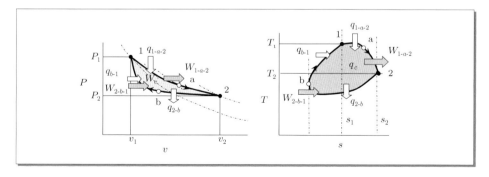

Fig.12-4　Cycle and entropy change for the case of Fig.12-2 (b)
　　　　　図 12-2(b) に対応したサイクルとエントロピー変化

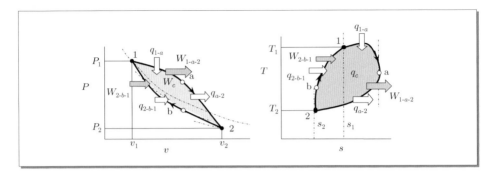

Fig.12-5   Cycle and entropy change for the case of Fig.12-2 (c)
図 12-2(c) に対応したサイクルとエントロピー変化

For the cases (b) and (c) shown in Fig.12-2, the position of state 2 corresponds to the maximum entropy or the minimum entropy as shown in Fig.12-4 and Fig.12-5.

It is important to note that the T-s chart is convenient for determining the direction of heat movement.   When a process is coupled with entropy increase, it means an endothermic change whatever it is expansion process or compression one.

図 12-2 における図 (b) や図 (c) の場合については, 図 12-4 や図 12-5 に示したように, 状態 2 の点はエントロピーの最大値か最小値のいずれかに対応する.

T-s 線図により, 熱の移動方向が簡単に理解可能になることは重要である. 変化過程が膨張変化過程であるか圧縮変化過程であるかにかかわらず, エントロピーの増加をともなう変化過程であれば, その変化過程は吸熱過程であることを示している.

12.3 Forward and Reverse Cycles

12.3 順サイクルと逆サイクル

Here, consider the direction of a cycle.   For a cycle shown with the P-v chart in Fig.12-6, it is considered that path "a" is positioned on the higher pressure level than path "b".   A cycle where processes proceed in the route 1-a-2-b-1 is called a forward cycle, and that proceeds through the route 1-b-2-a-1 is called a reverse cycle.   In other words, as shown in the P-v chart, the forward cycle is a clockwise cycle and the reverse cycle is a counterclockwise cycle.

For the forward cycle, the work done by the cycle given by Eq.12-11 is always positive.

サイクルの方向についてここで検討する. 図 12-6 の P-v 線図で示してあるサイクルについて, 経路 "a" は経路 "b" より高い圧力レベルにあるとする. この場合に 1-a-2-b-1 の順路で進行するサイクルを順サイクルとよび, 1-b-2-a-1 の順路で進行するサイクルを逆サイクルとよぶ. 言い換えると P-v 線図上において, 順サイクルは時計回りのサイクルであり, 逆サイクルは反時計回りのサイクルである.

順サイクルについては, 式 (12-11) で与えられるサイクル仕事はつねに正の値となる.

$$W_c = W_{1\text{-}a\text{-}2\text{-}b\text{-}1} = W_{1\text{-}a\text{-}2} + W_{2\text{-}b\text{-}1} = \oint_{1\text{-}a\text{-}2\text{-}b\text{-}1} Pdv \qquad (12\text{-}11)$$

$$W_c > 0 \quad (\text{positive work} \quad 正の仕事) \qquad (12\text{-}12)$$

Using the first law of thermodynamics and the definitions of the quantities of heat for q_a

熱力学の第一法則と, q_a や q_b の定義より, エネルギーバランスはつぎのようになる.

and q_b, the energy balance can be expressed as follows:

$$|q_a| = |q_b| + W_c \qquad (12\text{-}13)$$

Thus, the thermal efficiency for the forward cycle is always less than unity.

したがって，順サイクルの熱効率はいつも 1 より小さくなる.

$$\eta = \frac{W_c}{q_a} = \frac{|q_a| - |q_b|}{|q_a|} = 1 - \left|\frac{q_b}{q_a}\right| < 1 \qquad (12\text{-}14)$$

The reverse cycle is known as a refrigerator cycle or as a heat pump cycle. To complete the reverse cycle 1-b-2-a-1, the cycle receives a quantity of heat during the expansion process 1-b-2. For a compression process 2-a-1 of the reverse cycle, compression work is received and a quantity of heat has to be released. According to Eq.12-13, q_a is usually larger than q_b, and the average temperature level of the expansion process 1-b-2 is lower than the temperature in the process 2-a-1. Thus, the reverse cycle can act as a refrigerator cycle. The coefficient of refrigerator performance is defined by Eq.12-15.

逆サイクルは，冷凍サイクルやヒートポンプサイクルとして知られている．1-b-2-a-1 の逆サイクルを完結させるためには，1-b-2 の膨張変化過程の間にある熱量を受け取る必要がある．また，逆サイクルの 2-a-1 となる圧縮変化過程では，圧縮仕事を受け取り，ある熱量を放出することが必要になる．式(12-13)によれば，q_a はつねに q_b より大きく，さらに 1-b-2 の膨張変化過程の平均温度は 2-a-1 の変化過程の温度より低い．したがって，逆サイクルは冷凍サイクルとして作動する．冷凍サイクルの成績係数は，式（12-15）で定義されている．

$$[COP]_r = \frac{|q_b|}{|W_c|} = \frac{|q_b|}{|q_a| - |q_b|} > 1 \qquad (12\text{-}15)$$

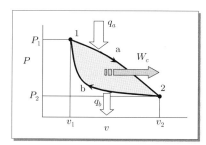

Fig.12-6 Forward cycle
順サイクル

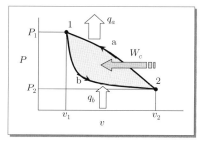

Fig,12-7 Reverse cycle
逆サイクル

Refrigerator utilizes an endothermic quantity of heat q_b, while heat pump utilizes an exothermic quantity of heat q_a. The coefficient of heat pump performance is given by Eq.12-16. Its value is higher than unity and also higher than $[COP]_r$.

吸熱される熱量 q_b は冷凍機で使われる. ヒートポンプでは, 放熱される熱量 q_a を利用する. ヒートポンプの成績係数は, 式(12-16)となる. この値は1より大きくなり, さらに $[COP]_r$ よりも大きくなる.

$$[COP]_h = \frac{|q_a|}{|W_c|} = \frac{|q_a|}{|q_a| - |q_b|} = 1 + \frac{|q_b|}{|q_a| - |q_b|} = 1 + [COP]_r > [COP]_r > 1 \qquad (12\text{-}16)$$

12.4 Reversible and Irreversible Cycles

12.4 可逆および非可逆サイクル

Reversible and irreversible cycles can be explained using Fig.12-8. First, consider a forward cycle that consists of reversible expansion and compression processes. The energy balance between the endothermic and exothermic quantities of heat and work is given by Eq.12-17. The total entropy change during a cycle is given by Eq.12-18. Since the cycle consists of only the reversible processes, no change in the total entropy results.

可逆サイクルと非可逆サイクルは, 図12-8を用いて説明することができる. まず, 可逆膨張過程と可逆圧縮過程から構成されている順サイクルを検討する. 吸熱および放熱の熱量と, 仕事の間に成り立つエネルギーバランスは, 式(12-17)で与えられる. サイクルの際のエントロピーの全体的な変化は, 式(12-18)で与えられている. サイクルは, 可逆プロセスのみから構成されているので, 全体としてはエントロピー変化が起こらない.

$$|q_a| = |q_b| + W_c \qquad (12\text{-}17)$$

$$\Delta s = s_{1-a-2-b-1} = \int_{1-a}^{2} \frac{dq_a}{T} + \int_{2-b}^{1} \frac{dq_b}{T} = \oint_{1-a-2-b-1} \frac{dq}{T} = 0 \qquad (12\text{-}18)$$

No entropy change for all the processes in a reversible cycle is one of the important results of the thermodynamics. Since entropy is one of

可逆サイクルの全変化過程に対しては, エントロピーの変化がなく, このことは, 熱力学としてきわめて重要な結果の一つである.

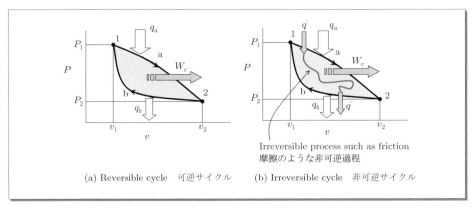

(a) Reversible cycle   可逆サイクル   (b) Irreversible cycle   非可逆サイクル

Fig.12-8   Reversible and irreversible cycles   可逆サイクルと非可逆サイクル

the state quantities that is defined uniquely by two other state quantities such as pressure and temperature, entropy change between states 1 and 2 should be uniquely defined even if the processes such as 1-a-2 and 1-b-2 are different. Thus, no entropy change as shown by Eq.12-18 reasonably expresses the nature of a reversible cycle.

The reversible processes of the cycle are shown in Fig.12-8 (a). However, when an irreversible process such as friction is included in the cycle, the quantity of heat q' corresponding to this irreversible process is additionally received by the cycle and directly released from the cycle as shown in Fig.12-8 (b). For this case, the energy balance can be noted as follows.

エントロピーは状態量の一つであり，圧力は温度のような他の二つの状態量により状態に対応した一つの値として定まるので，1-a-2 や 1-b-2 ように変化過程が異なるとしても，状態 1 と状態 2 の間のエントロピー変化は一つの値となるべきものである．したがって，式（12-18）で示したエントロピー変化が現れないことは，可逆サイクルの性質として理に適っている内容であるといえる．

サイクルのこの可逆変化は，図 12-8（a）に示されている．ところが，摩擦のような非可逆変化がサイクルに含まれている場合では，非可逆過程に対応する熱量 q' がサイクルに付加的に受け取られ，かつそれが直接的にサイクルから放出される．この状況は，図 12-8（b）に示したとおりである．この場合，エネルギーバランスは以下のようになる．

$$|q_a|+|q'|=|q_b|+|q'|+W \tag{12-19}$$

Similar to the case of Fig.12-8 (a), the entropy change of this system is noted by Eq.12-20. Since no entropy change results in a reversible cycle, it can be reduce to Eq.12-21.

このシステムのエントロピー変化は，図 12-8（a）の場合と同様に，式（12-20）のように記述される．さらに，可逆サイクルではエントロピー変化が起きないことから，この式は式（12-21）のようになる．

$$\Delta s' = s_{1-a-2-b-1} = \int_{1-a}^{2} \frac{dq_a}{T} + \int_{1-a}^{2} \frac{dq'}{T} + \int_{2-b}^{1} \frac{dq_b}{T} + \int_{2-b}^{1} \frac{dq'}{T}$$
$$= \oint_{1-a-2-b-1} \frac{dq}{T} + \int_{1-a}^{2} \frac{dq'}{T} + \int_{2-b}^{1} \frac{dq'}{T} \tag{12-20}$$

$$\Delta s' = \int_{1-a}^{2} \frac{dq'}{T} + \int_{2-b}^{1} \frac{dq'}{T} \tag{12-21}$$

Considering the endothermic and exothermic heat movements for the cycle and both temperatures of paths "a" and "b", it is clear that the entropy change for an irreversible cycle is negative.

吸熱および放熱という熱の動きをサイクルについて検討し，また，経路 "a" と "b" についての両者の温度を検討すれば，非可逆サイクルについてのエントロピー変化が負であることが明らかになる．

$$\int_{1-a}^{2} \frac{dq'}{T} + \int_{2-b}^{1} \frac{dq'}{T} = \int_{1-a}^{2} \frac{dq'}{T} - \int_{1-b}^{2} \frac{dq'}{T} < 0 \tag{12-22}$$

$$\Delta s' < 0 \tag{12-23}$$

This means that the entropy released from the irreversible cycle is larger than the entropy received. In other words, an irreversible cycle re-

このことは，非可逆サイクルから放出されるエントロピーはそのサイクルが受け取るエントロピーより大きいことを示している．言

sults in an increase of entropy.

い換えれば，非可逆サイクルはエントロピーの増加を引き起こしている．

△ 12-1

A thermodynamic cycle receives 500 kJ of heat from a high temperature heat source and rejects 320 kJ of heat to a low temperature heat sink. Determine the work produced by this cycle and the thermal efficiency of the cycle.

△ 12-2

Consider a refrigerator in which the coefficient of refrigerator performance [COP], is 2.50. If power of 5 kJ is added to this refrigerator, determine the amount of heat removed from the refrigerated space.

△ 12-3

A reversible cycle receives 300 J/K of entropy from a high temperature heat source at 1 000 K, and rejects 120 kJ of waste heat to a low temperature heat sink. Determine (1) the amount of heat received from the heat source, (2) the temperature of the heat sink, and (3) the thermal efficiency of the cycle.

△ 12-1

ある熱力学的サイクルが，高温熱源から 500 kJ の熱を受け取り，低温熱源に 320 kJ の熱を排出した．このときこのサイクルがした仕事と，サイクルの熱効率を求めなさい．

△ 12-2

成績係数が 2.50 の冷凍機を考える．5 kJ の動力がこの冷凍機に入力されるとき，冷凍庫から取り除かれる熱量を求めなさい．

△ 12-3

ある可逆サイクルが，1 000 K の高温熱源から 300 J/K のエントロピーを受け取り，低温熱源に 120 kJ の熱を放出する．このとき，(1) 高温熱源から受け取る熱量，(2) 低温熱源の温度，(3) このサイクルの熱効率をそれぞれ求めなさい．

Chapter 13

Carnot Cycle and Entropy Change in Isolated System
カルノーサイクルと孤立系のエントロピー変化

13.1 Carnot Cycle

A special cycle called "Carnot cycle", in which the working media is an ideal gas, is considered. This cycle consists of two isothermal changes for both heat received and heat released, and two adiabatic changes. In other words, as shown in Fig.13-1, the expansion process consists of an isothermal expansion and an adiabatic expansion; the compression process consists of an isothermal compression and an adiabatic compression.

According to the notations indicated in Fig.13-1, an isothermal change from state 1 to 2 is specified by Eq.13-1.

13.1 カルノーサイクル

理想気体を作動気体とするカルノーサイクルという特別なサイクルを検討する. このサイクルは, 熱を受け入れる変化と, 熱を放出するという二つの等温変化と二つの断熱変化から構成されている. 言い換えると, 図 13-1 に示してあるように, 膨張変化過程が等温膨張と断熱膨張から構成され, 圧縮変化過程が等温圧縮と断熱圧縮から構成されている.

図 13-1 に示した表記に従えば, 状態 1 から状態 2 への等温変化は, 式 (13-1) により確定されている.

$$T_a = T_1 = T_2, \quad Pv = P_1 v_1 = P_2 v_2 = \text{const.} \tag{13-1}$$

During this change, the cycle receives heat and produces expansion work. Following this change, further expansion work is done by an adiabatic expansion change (change from 2 to 3) as specified by Eq.13-2.

この変化の間, カルノーサイクルは一定の熱量を受け取り, また, 膨張仕事を行う. この変化に引き続き, 断熱変化 (2 から 3 への変化) によりさらに膨張仕事が行われ, これは式 (13-2) で定められている.

$$Pv^\kappa = P_2 v_2^\kappa = P_3 v_3^\kappa = \text{const.}, \quad Tv^{\kappa-1} = T_2 v_2^{\kappa-1} = T_3 v_3^{\kappa-1} = \text{const.},$$

$$\frac{P^{\frac{\kappa-1}{\kappa}}}{T} = \frac{P_2^{\frac{\kappa-1}{\kappa}}}{T_2} = \frac{P_3^{\frac{\kappa-1}{\kappa}}}{T_3} = \text{const.} \tag{13-2}$$

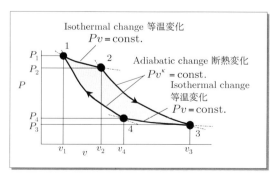

Fig.13-1 Carnot cycle
カルノーサイクル

Compression process is also divided into two changes. One is an isothermal compression change (Eq.13-3) from state 3 to 4 in which a certain quantity of heat is released.

圧縮変化過程もまた，二つの変化に分けられている．その一つは，状態3から状態4への等温変化（式（13-3））であり，この変化により一定量の熱がサイクルから放出される．

$$T_b = T_3 = T_4, \quad Pv = P_3 v_3 = P_4 v_4 = \text{const.} \tag{13-3}$$

The other is the final adiabatic change (Eq.13-4) from state 4 to 1 in which the thermodynamic state of the working gas returns to its initial state.

残りの変化過程は，状態4から状態1への最終的な断熱変化（式（13-4））であり，作動気体の熱力学的な状態は初期状態にもどることになる．

$$Pv^{\kappa} = P_4 v_4^{\kappa} = P_1 v_1^{\kappa} = \text{const.}, \quad Tv^{\kappa-1} = T_4 v_4^{\kappa-1} = T_1 v_1^{\kappa-1} = \text{const.},$$
$$\frac{P^{\frac{\kappa-1}{\kappa}}}{T} = \frac{P_4^{\frac{\kappa-1}{\kappa}}}{T_4} = \frac{P_1^{\frac{\kappa-1}{\kappa}}}{T_1} = \text{const.} \tag{13-4}$$

13.2 Thermal Efficiency of Carnot Cycle

13.2 カルノーサイクルの熱効率

The Carnot cycle is simply expressed on a T-s chart because this cycle is specified by the process equations: $dT = 0$ and $ds = 0$. As shown in Fig.13-2, the area that is enclosed by the Carnot cycle (1-2-3-4) in the T-s chart is equal to the actual quantity of heat that is consumed in the cycle.

カルノーサイクルは，$dT = 0$ と $ds = 0$ という変化過程の特性方程式で記述されているので，T-s 線図上にわかりやすく表現することができる．図 13-2 に示したとおり，1-2-3-4 のカルノーサイクルで囲まれた T-s 線図の面積は，サイクルで消費する実質的な熱量に一致している．

The quantity of heat received during the isothermal change from state 1 to 2 is simply given by Eq.13-5.

状態1から状態2への等温変化の間にサイクルが受け取った熱量は，式（13-5）で与えられることが明らかである．

$$q_{1-2} = \int_1^2 T ds = (s_2 - s_1) T_a \tag{13-5}$$

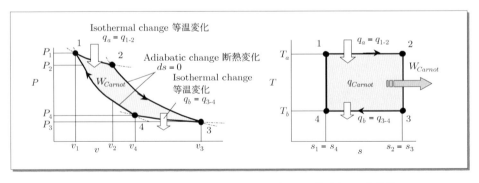

Fig.13-2 Thermal efficiency of carnot cycle   カルノーサイクルの熱効率

Similarly, the quantity of heat released during the isothermal change from state 3 to 4 is given by Eq.13-6.

同様に，状態 3 から状態 4 への等温変化の間にサイクルが放出する熱量は，式（13-6）で与えられる．

$$q_{3-4} = \int_3^4 Tds = (s_4 - s_3)T_b = -(s_2 - s_1)T_b \qquad (13\text{-}6)$$

From the first law of thermodynamics, the actual work done by the cycle is given by the difference between the quantities of heat received and heat released. Furthermore, it can be expressed using the entropy terms of Eq.13-5 and Eq.13-6. Then, the work is given by Eq.13-7.

熱力学の第一法則より，サイクルによって行われる実質的な仕事は，サイクルが受け取った熱とサイクルが放出した熱の差で与えられている．さらに，これは式（13-5）と式（13-6）のエントロピーの項を用いて表すことが可能である．したがって，仕事は式（13-7）で与えられる．

$$W_{Carnot} = |q_a| - |q_b| = (s_2 - s_1)(T_a - T_b) \qquad (13\text{-}7)$$

From the definition of thermal efficiency of a cycle, the Carnot cycle thermal efficiency is derived as follows:

サイクルの熱効率の定義から，カルノーサイクルの熱効率はつぎのように導かれる．

$$\eta_{Carnot} = \frac{W_{Carnot}}{q_a} = \frac{(s_2 - s_1)(T_a - T_b)}{(s_2 - s_1)T_a} = 1 - \frac{T_b}{T_a} \qquad (13\text{-}8)$$

From this derivation (Eq.13-8), it is clear that the efficiency depends only on the two isothermal temperatures in the cycle.

この熱効率は，カルノーサイクルの二つの等温変化における温度にのみ従属していることが式（13-8）より明らかである．

13.3 Maximum Thermal Efficiency of Heat Engines

13.3 熱機関の最大効率

A heat engine has its own thermodynamic cycle. When the cycle is illustrated on a T-s chart, it is characterized by the maximum and minimum temperatures and by the entropies of the maximum and minimum quantities. In Fig.13-3, the heat engine cycle is specified by

熱機関は，それぞれ固有の熱力学的サイクルを有している．T-s 線図上にサイクルを表現すれば，サイクルは最高温度と最低温度という二つの温度および，最大と最低を示すエントロピーの二つの量によって特徴づけられている．図 13-3 において，熱機関のサイク

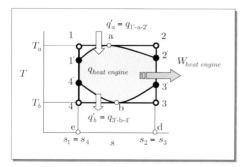

Fig.13-3　T-s chart for a heat engine
　　　　　熱機関の T-s 線図

the cycle 1'-a-2'-3'-b-4', and the quantity of heat consumed in this cycle is given by the enclosed area 1'-a-2'-3'-b-4'.

The quantity of heat q'_a that is received by this heat cycle is less than the quantity of heat q_a that is received by the Carnot cycle shown in Fig.13-2. The difference between the quantities of heat received by both cycles is given by Eq.13-9.

ルは 1'-a-2'-3'-b-4' のサイクルとなり，サイクルで消費する熱量は 1'-a-2'-3'-b-4' で囲まれた T-s 線図上の面積となる．

この熱サイクルに受け取られる熱量 q'_a は，図 13-2 に示したカルノーサイクルが受け取る熱量 q_a より少なくなる．両方のサイクルで受け取る熱の差については，式（13-9）で与えられる．

$$\Delta q_a = q_a - q'_a = \left[\text{area 1-2-d-e-1} \right] - \left[\text{area 1'-a-2'-d-e-1'} \right] \tag{13-9}$$

As for the quantities of heat released, their difference is obtained as follows:

放出する熱については，両者の差が次式のように得られる．

$$\Delta q_b = q'_b - q_b = \left[\text{area 4'-b-3'-d-e-4'} \right] - \left[\text{area 4-3-d-e-4} \right] \tag{13-10}$$

From the definition of thermal efficiency, the heat engine's thermal efficiency can be obtained as follows:

熱効率の定義より，熱機関の熱効率は以下のようになる．

$$\eta_{heat\ engine} = \frac{q'_a - q'_b}{q'_a} = \frac{(|q_a| - |\Delta q_a|) - (|q_b| + |\Delta q_b|)}{(|q_a| - |\Delta q_a|)} = 1 - \frac{(|q_b| + |\Delta q_b|)}{(|q_a| - |\Delta q_a|)} \leq 1 - \frac{|q_b|}{|q_a|} \tag{13-11}$$

$$\eta_{heat\ engine} \leq 1 - \frac{|q_b|}{|q_a|} = 1 - \frac{T_b}{T_a} = \eta_{\text{Carnot}} = \eta_{th} \tag{13-12}$$

Equation 13-12 indicates that the efficiency of the heat engine is less than that of the Carnot cycle working under temperature conditions of T_a and T_b. In other words, the Carnot cycles gives the maximum thermal efficiency of cycles working under the same limited temperature conditions.

式（13-12）は，熱機関の熱効率は T_a と T_b の温度条件下で作動するカルノーサイクルの熱効率より小さいことを示している．いいかえれば，同じ温度制限下で作動する熱機関の最大効率をカルノーサイクルは与えていることになる．

13.4 Clausius' Integral

13.4 クラウジウスの積分

Consider a thermodynamic reversible cycle shown in Fig.13-4. As shown in this figure, it is assumed that this cycle can be replaced with many small Carnot cycles. The Carnot cycle of the i-th number consists of isothermal change from state 1 to 2, adiabatic change from 2 to 3, isothermal change from 3 to 4, and a final adiabatic change from state 4 to state 1.

The quantity of heat supplied during the isothermal change from state 1 to state 2 is expressed as follows:

熱力学的な可逆サイクルを検討する．図 13-4 に示したように，このサイクルは多くの小さなカルノーサイクルで置き換えられると想定する．i 番目のカルノーサイクルは，状態 1 から状態 2 への等温変化，状態 2 から状態 3 への断熱変化，状態 3 から状態 4 への等温変化，そして状態 4 から状態 1 への最終的な断熱変化から構成されている．

状態 1 から状態 2 への等温変化の間に供給された熱量は，つぎのように表される．

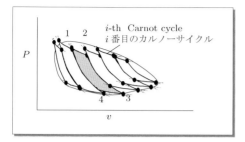

Fig.13-4 Multi-cycle replacement using carnot cycle
カルノーサイクルを使った多サイクル置換

$$q_a = \int_1^2 Pdv = RT_a \int_1^2 \frac{dv}{v} = RT_a \ln\frac{v_2}{v_1} = RT_a \ln\frac{P_1}{P_2} \tag{13-13}$$

Also, the quantity of heat released during the change from state 3 to 4 is similarly given by Eq.13-14.

また，状態3から状態4への変化のあいだに放出した熱量は，式（13-14）により同様に与えられる．

$$q_b = \int_3^4 Pdv = RT_b \int_3^4 \frac{dv}{v} = RT_b \ln\frac{v_4}{v_3} = RT_b \ln\frac{P_3}{P_4} \tag{13-14}$$

Considering an adiabatic condition for the changes from state 2 to 3 and from state 4 to 1, the pressure ratios for these changes are given by Eqs.13-15 and 13-16.

状態2から状態3，また状態4から状態1への変化に対する断熱条件により，断熱変化での圧力比は，式（13-15）と式（13-16）で与えられる．

$$\frac{P_1}{P_4} = \left(\frac{v_4}{v_1}\right)^\kappa = \left(\frac{T_a}{T_b}\right)^{\frac{\kappa}{\kappa-1}} \tag{13-15}$$

$$\frac{P_3}{P_2} = \left(\frac{v_2}{v_3}\right)^\kappa = \left(\frac{T_b}{T_a}\right)^{\frac{\kappa}{\kappa-1}} \tag{13-16}$$

Then, the entropy change for the i-th Carnot cycle expressed by Eq.13-17 becomes zero.

したがって，式（13-17）で表現されているi番目のカルノーサイクルのエントロピー変化は，ゼロとなる．

$$\frac{q_a}{T_a} + \frac{q_b}{T_b} = R\left(\ln\frac{P_1}{P_2} + \ln\frac{P_3}{P_4}\right) = R\ln\frac{P_1P_3}{P_2P_4} = 0 \tag{13-17}$$

This is another verification of no entropy change for the Carnot cycle shown in Eq.12-18. Here, q_b has a negative value which corresponds to the quantity of heat released from the i-th Carnot cycle.

Since the above analysis is made for the i-th Carnot cycle and since the whole cycle can be replaced with a summation of all the small Carnot cycles, the entropy change for the whole cy-

これは，式（12-18）に示したカルノーサイクルにおいてエントロピー変化がないことの別の証明である．ここで，q_bは負の値をもち，これはi番目のカルノーサイクルからの熱の放出量に対応している．

上記の解析は，i番目のサイクルに対応したものであり，全体のサイクルは小さなカルノーサイクルの集合により置き換えることができるから，全体のサイクルのエントロピー

cle can be expressed using Eq.13-18 or Eq.13-19.

変化は式（13-18）または式（13-19）で表すことができる.

$$\sum_i \frac{q_i}{T_i} = 0 \qquad (13\text{-}18)$$

$$\oint \frac{dq}{T} = 0 \qquad (13\text{-}19)$$

The integral equation shown by Eq.13-19 is called the "Clausius' integral". The whole cycle considered here is a reversible cycle. However the above analysis can be applicable to any kind of reversible cycle. Then, Eq.13-20 can be used as a general expression for reversible cycles.

式（13-19）に示した積分が，"クラウジウスの積分"とよばれているものである. ここで検討しているサイクル全体は可逆サイクルであるが，上記の解析は，任意の可逆サイクルに対して適応することができる. したがって，可逆サイクルに対する一般的な表現としては，式（13-20）を導くことができる.

$$\Delta s_{\substack{reversible \\ cycle}} = \int_{\substack{reversible \\ cycle}} ds = \oint_{\substack{reversible \\ cycle}} \frac{dq}{T} = 0 \qquad (13\text{-}20)$$

As for an irreversible cycle, extra quantities of heat are released. Following this fact, an entropy change for an irreversible cycle can be expressed by Eq.13-21and Eq.13-22. Eq.13-22 means that extra entropy is produced and released from an irreversible cycle.

非可逆サイクルでは，サイクルに対して余分な熱量の放出がある. この事実に従えば，非可逆サイクルのエントロピー変化は，式（13-21）や式（13-22）で表現することができる. 式（13-22）は，余分のエントロピーが非可逆サイクルで生じ，それがサイクルから放出されることを示している.

$$\Delta s_{\substack{irreversible \\ cycle}} = \oint_{\substack{irreversible \\ cycle}} ds = \oint_{\substack{reversible \\ cycle}} \frac{dq}{T} - \int_{\substack{irreversible \\ process}} \frac{|q_{loss}|}{T} < 0 \qquad (13\text{-}21)$$

$$\oint_{\substack{irreversible \\ cycle}} \frac{dq}{T} < \oint_{\substack{reversible \\ cycle}} \frac{dq}{T} = 0 \qquad (13\text{-}22)$$

13.5 Entropy Change for an Isolated System

13.5 孤立系のエントロピー変化

An isolated system is a thermodynamic system that contains various heat sources and various thermodynamic cycles but is thermally and kinetically isolated from its surroundings. Here, using Fig.13-5, we consider an isolated system comprising a Carnot cycle.

It consists of high and low temperature heat sources of which the temperatures are T_H and T_L respectively. The Carnot cycle works between a high temperature T_a and a low tempera-

種々の熱源や，多くの熱力学的サイクルを包含し，かつ外界からは熱的および力学的に断絶している熱力学的システムを孤立系という. ここで，図13-5を用いてカルノーサイクルを内部に包含する孤立系を検討する.

この孤立系は，温度がそれぞれ T_H と T_L となる高温と低温の熱源，および高温 T_a と低温 T_b の間で作動しているカルノーサイクルから構成されている. 高温熱源からカル

ture T_b. Assuming a quasi-static heat transportation process from the high temperature heat
source to the Carnot cycle, the quantity of heat
q_a indicates to both the quantity of heat released
from the source and the quantity of heat received during the isothermal expansion process
of the Carnot cycle.

Also, the temperatures T_H and T_a become the
same because equal since heat transportation is
quasi-static. As for the Carnot cycle, the first
law of thermodynamics results in the energy
balance shown by Eq.13-23.

ノーサイクルへの熱の移動が準静的変化であ
るとすれば，熱量 q_a は，熱源から放出され
た熱量とカルノーサイクルの等温膨張変化過
程で受け取られた熱量の両者を示している．

また，準静的変化過程であるので，熱源温
度 T_H とカルノーサイクルの温度 T_a は等しく
なる．カルノーサイクルについては，熱力学
の第一法則が式（13-23）に示してあるエネ
ルギーバランスを与えている．

$$|q_a|=|q_b|+W_{Carnot} \tag{13-23}$$

Since the Carnot cycle is a reversible cycle,
no entropy change is expected. This is shown by
Eq.13-24.

カルノーサイクルは，可逆サイクルの一つ
であり，エントロピーの変化はないと考える
ことができるので，式（13-24）が成り立つ．

$$\frac{|q_a|}{T_a}=\frac{|q_b|}{T_b} \tag{13-24}$$

Here, q_b is the quantity of heat released from
the Carnot cycle and is the same quantity of heat
received by the low temperature heat source.
The assumption of a quasi-static heat transporta-

ここで，q_b はカルノーサイクルから放出さ
れる熱量であり，同じ熱量が低温熱源に受け
取られることになる．カルノーサイクルの等
温圧縮変化過程から，低温熱源へ熱が準静的

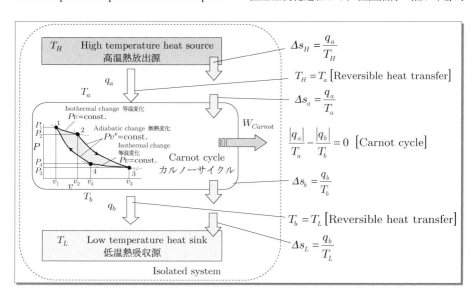

Fig.13-5　Isolated system comprising a Carnot cycle　カルノーサイクルを包含する孤立系

tion process from the isothermal compression process to the low temperature heat source, T_L should be the same with T_b. Then, the following equations are derived for the isolated system that contains quasi-static heat transportation and reversible cycle such as the Carnot cycle.

に移動するとしているので，T_L は T_b と同じになる．したがって，準静的は熱移動とカルノーサイクルような可逆サイクルを含む孤立系については，以下のような関係が成立する．

$$\frac{|q_a|}{T_H} = \frac{|q_a|}{T_a} = \frac{|q_b|}{T_b} = \frac{|q_b|}{T_L} \tag{13-25}$$

$$\Delta s_{isolated\ system} = \frac{|q_b|}{T_L} - \frac{|q_a|}{T_H} = 0 \tag{13-26}$$

The temperature difference between T_H and T_a is needed for actual transportation of heat, and in a real thermodynamic cycle heat loss such as friction loss always exists. This means that a real isolated system includes various kinds of irreversible thermodynamic processes.

Considering the temperature difference needed for heat transportation and the irreversible heat loss in a cycle, the entropy change of a typical isolated system can be derived as shown in

熱の実際の移動においては，T_H と T_a の間に温度差が必要である．また，摩擦損失のような熱損失が，実際の熱力学サイクルの内部には存在している．このことは，実際の孤立系は多くの非可逆的な熱力学的変化過程を包含していることを意味している．

熱の移動のための温度差や，サイクル中の非可逆な熱損失を検討することにより，一般的な孤立系のエントロピー変化を図 13-6 に示したように導きだすことができる．

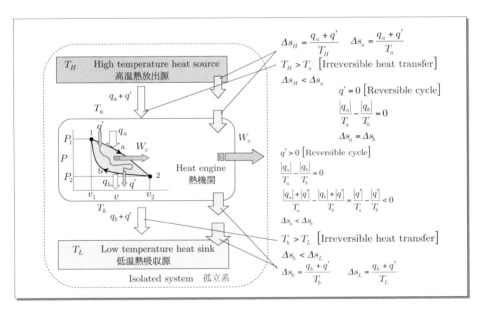

Fig.13-6   Isolated system including irreversible processes
非可逆変化過程を包含する孤立系

Fig.13-6.

Equation 13-27 shows that entropy increases during each irreversible process. Eq.13-28 shows the general entropy increase of an isolated system.

式（13-27）は，それぞれの非可逆変化過程の間にエントロピーが増大することを示している．また，式（13-28）は，孤立系の一般的なエントロピー増大を示している．

$$\frac{|q_a|+|q'|}{T_H} < \frac{|q_a|+|q'|}{T_a} \le \frac{|q_b|+|q'|}{T_b} < \frac{|q_b|+|q'|}{T_L} \tag{13-27}$$

$$\Delta s_{isolated\ system} = \frac{|q_b|+|q'|}{T_L} - \frac{|q_a|+|q'|}{T_H} > 0 \tag{13-28}$$

13.6 Exergy

Consider a cascade energy utilization model indicated in Fig.13-7. A quantity of heat that is equivalent to the internal energy of 1 kg mass of an ideal gas is assumed to be released with its heat carrier from a high temperature heat source.

When the heat carrier gives a part of its internal energy to a Carnot cycle, its temperature level decreases. This means that a Carnot cycle working under a fixed isothermal expansion process can not receive all of the internal energy of the carrier. Thus, many Carnot cycles of different temperature levels are needed to utilize all of the internal energy of the carrier. In order to receive all of the heat energy, a cascade system

13.6 エクセルギー

図 13-7 に示したカスケード形のエネルギー利用モデルを検討する．理想気体 1 kg の内部エネルギー相当の熱が，その熱媒体とともに高温熱源から放出されることを想定する．

熱媒体がその内部エネルギーの一部をカルノーサイクルに与えれば，熱媒体の温度は低下する．このことは，一定の等温膨張変化過程で作動するカルノーサイクルは熱媒体の内部エネルギーのすべてを受けることができないことを意味している．したがって，熱媒体の内部エネルギーのすべてを利用するためには，温度レベルの異なる多段のカルノーサイクルが必要になる．熱エネルギーをすべて受

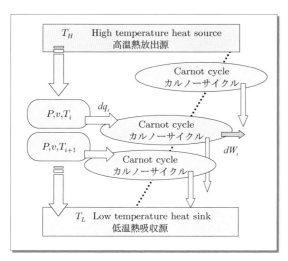

Fig.13-7 Cascade utilization of heat energy
　　　　　熱エネルギーのカスケード利
　　　　　用

of Carnot cycles is considered. In this system, the i-th Carnot cycle consists of an isothermal expansion process of temperature T_i to receive heat and an isothermal compression process of temperature T_L to release heat.

The quantity of heat that can be received by the i-th Carnot cycle can be given by Eq.13-29.

け取るために，ここではカルノーサイクルのカスケードシステムを想定する．このシステムにおいて，i番目のカルノーサイクルは，熱を受け取るために温度T_iの等温膨張変化過程と，熱の放出のための温度T_Lの等温圧縮過程を有している．

i番目のカルノーサイクルが受け取ることのできる熱量は，式（13-29）で与えられる．

$$dq_i = c_v(T_{i+1} - T_i) = -c_v dT_i \tag{13-29}$$

The i-th Carnot cycle produces work and it is given by Eq.13-30.

i番目のカルノーサイクルが仕事を生みだし，それが式（13-30）で与えられている．

$$dW_i = \eta_{i\text{-}th\,Carnot} dq_i = -c_v\left(1 - \frac{T_L}{T_i}\right)dT_i \tag{13-30}$$

Then, the total work produced from the quantity of heat that is carried by the heat carrier is obtained as follows:

したがって，熱媒体によって運ばれている熱量から生じる仕事の全体は，以下のようになる．

$$W_{H-L} = -c_v\int_H^L\left(1 - \frac{T_L}{T_i}\right)dT = c_v\int_L^H dT - c_v T_L \int_L^H \frac{dT}{T} \tag{13-31}$$

$$W_{H-L} = c_v(T_H - T_L) - c_v T_L \ln\frac{T_H}{T_L} \tag{13-32}$$

$$e_{exergy} = W_{H-L} \tag{13-33}$$

This work gives the maximum work that can be converted from the internal energy of the carrier. It is called the "exergy" or effective energy in thermodynamics.

There is no work done by the carrier itself during the above processes, and changes in internal energy and entropy of the carrier are as follows:

この仕事は，熱媒体の内部エネルギーから変換可能な最大の仕事を与えている．熱力学ではこれをエクセルギー，または有効エネルギーとよんでいる．

上記の変化過程の間に熱媒体自体による仕事は行われていないので，熱媒体の内部エネルギーやエントロピーの変化はつぎのようになる．

$$\Delta u_{L-H} = c_v(T_H - T_L) \tag{13-34}$$

$$\Delta s_{L-H} = \int_L^H \frac{dq}{T} = c_v\int_L^H \frac{dT}{T} = c_v \ln\frac{T_H}{T_L} \tag{13-35}$$

Then, the exergy can be given by Eq.13-36.

したがって，エクセルギーは式（13-36）で与えられる．

$$e_{exergy} = \Delta u_{L-H} - T_L \Delta s_{L-H} \tag{13-36}$$

Furthermore, when the internal energy and

さらに，内部エネルギーとエントロピーの

entropy at the low temperature level are zero, exergy takes the definition of "Helmholtz's free energy" introducced by Eq.6-37 and can be expressed using the efficiency of a Carnot cycle.

低温熱源のレベルをゼロとすれば，エクセルギーは，式（6-37）で導入したヘルムホルツの自由エネルギーの定義と一致する．さらに，エクセルギーはカルノーサイクルの効率を用いて表現することができる．

$$u_L = 0, \quad s_L = 0 \tag{13-37}$$

$$e_{\mathrm{exergy}} = u_H - T_L s_H \qquad (\text{Helmholtz's free energy}) \tag{13-38}$$

$$e_{\mathrm{exergy}} = u_H - T_L \frac{g_H}{T_H} = u_H \left(1 - \frac{T_L \dfrac{c_v T_H}{T_H}}{c_v T_H} \right) = \eta_{Carnot(H-L)} u_H \tag{13-39}$$

Exergy gives the maximum energy that is convertible to the work from the internal energy of the heat carrier. As for the more general conclusion, the maximum utilized (available) energy of the high temperature heat source is limited by the efficiency of the Carnot cycle that is operated between high and low temperature heat sources.

エクセルギーは，熱媒体のもつ内部エネルギーから仕事と変換可能な最大のエネルギーを与えている．一般的にいえば，高温熱源のもつ最大有効エネルギーは，高温と低温の両熱源間にて作動するカルノーサイクルの効率によって規定されているということである．

Problems

△13-1

Consider that a reversible Carnot cycle works with a high temperature heat source at 1000 K and a low temperature heat sink at 300 K. If it receives 300 kJ per 1 kg of the working fluid from a heat source, determine (1) the thermal efficiency of this cycle, (2) the work per 1 kg of the working fluid produced by this cycle and (3) the amount of heat per 1 kg of the working fluid rejected to the low temperature heat sink.

△13-2

Consider that a reversible Carnot cycle works with a high temperature heat source at 1000 K and a low temperature heat sink at 400 K. If this cycle produces 0.48 kJ of work, determine (1) the amount of heat received from the heat source and rejected to the heat sink, and (2) the change in entropy in the cycle.

問題

△13-1

1 000 K の高温熱源と 300 K の低温熱源で作動する可逆カルノーサイクルを考える．このサイクルが作動流体 1 kg あたり 300 kJ の熱量を高温熱源から受け取るとき，（1）このサイクルの熱効率，（2）このサイクルで発生する作動流体 1 kg あたりの仕事，（3）作動流体 1 kg あたりに低温熱源に排出する熱量をそれぞれ求めなさい．

△13-2

1 000 K の高温熱源と 400 K の低温熱源で作動する可逆カルノーサイクルを考える．このサイクルが 0.48 kJ の仕事を発生するとき，（1）高温熱源から受け取る熱量と低温熱源に排出する熱量，（2）このサイクルにおけるエントロピー変化をそれぞれ求めなさい．

△13-3

Consider a reversible Carnot cycle shown in Fig.13-1. If q_a = 40 kJ/kg, T_a = 1000 K, q_b = 16 kJ/kg, and P_3 = 0.05 MPa, determine T_b and P_4. Here, the working fluid is an ideal gas and the gas constant is 0.287 kJ/(kg·K).

△13-3

Fig.13-8 のような可逆カルノーサイクルを考える。ここで，q_a = 40 kJ/kg, T_a = 1 000 K, q_b = 16 kJ/kg, P_3 = 0.05 MPa とするとき，T_b と P_4 を求めなさい．作動流体は理想気体でガス定数は 0.287 kJ/（kg·K）である．

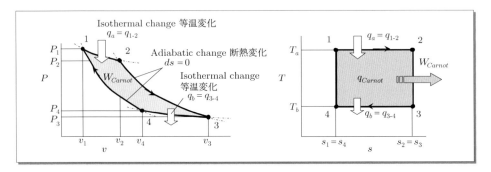

Fig.13-8   Reversible Carnot cycle for Problem 13-3
問題 13-3 の可逆カルノーサイクル

Chapter 14
Gas Cycle and Heat Engine
ガスサイクルと熱機関

14.1 Stirling Cycle

14.1 スターリングサイクル

Consider a unique cycle that consists of an isothermal expansion process to receive heat and work, a constant volume cooling for pressure reduction with heat release, an isothermal compression to release heat and negative work, and a final constant volume heating to receive heat. This cycle is called a "Stirling cycle". The $P\text{-}v$ and $T\text{-}s$ charts of this cycle are shown in Fig.14-1.

熱の受領と仕事のための等温膨張変化過程，圧力低下と熱の放出のための等容冷却変化過程，熱の放出と負の仕事のための等温圧縮変化過程，それから熱の受領のための等容加熱変化過程からなる独特なサイクルを検討する．このサイクルは，スターリングサイクルとよばれている．このサイクルの $P\text{-}v$ 線図と $T\text{-}s$ 線図が図 14-1 に示してある．

Thermodynamic quantities during the isothermal expansion from state 1 to 2 are given by Eq.14-1.

状態 1 から状態 2 への等温変化の間の熱力学的諸量は式（14-1）のように与えられる．

$$T_a = T_1 = T_2, \quad Pv = P_1 v_1 = P_2 v_2 = \text{const.}, \quad W_{1-2} = q_{1-2} = q_a \tag{14-1}$$

Since the process is isothermal, the positive expansion work is equivalent to the quantity of heat received. The thermodynamic quantities during the cooling process from state 2 to 3 are obtained by Eq.14-2.

ここでの変化過程は等温であるので，正の膨張仕事は変化過程で受領した熱量に等しい．状態 2 から状態 3 への冷却変化過程での熱力学的諸量は，式（14-2）で得られる．

$$v = v_2 = v_3 = \text{const} \;\Rightarrow\; q_{2-3} = c_v (T_3 - T_2) = c_v (T_b - T_a),$$
$$s_{2-3} = s_3 - s_2 = c_v \ln \frac{T_3}{T_2} = c_v \ln \frac{T_b}{T_a} \tag{14-2}$$

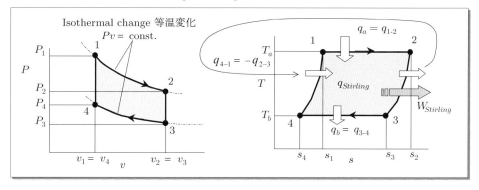

Fig.14-1 $P\text{-}v$ and $T\text{-}s$ charts of Stirling cycle　スターリングサイクルの $P\text{-}v$ 線図と $T\text{-}s$ 線図

As for the isothermal compression process from state 3 to 4, Eq.14-3 is obtained.

状態3から状態4への等温圧縮に関しては，式（14-3）が得られる．

$$T_b = T_3 = T_4, \quad Pv = P_3 v_3 = P_4 v_4 = \text{const.}, \quad W_{3-4} = q_{3-4} = q_b \tag{14-3}$$

The quantity of heat received during the final heating process from state 4 to 1, and the entropy change are given by Eq.14-4.

状態4から状態1への最終的な加熱過程の間に受領した熱量とエントロピー変化は，式（14-4）で与えられる．

$$v = v_4 = v_1 = \text{const.} \quad \Rightarrow \quad q_{4-1} = c_v (T_1 - T_4) = c_v (T_b - T_a),$$
$$s_{4-1} = s_1 - s_4 = c_v \ln \frac{T_1}{T_4} = c_v \ln \frac{T_a}{T_b} \tag{14-4}$$

From Eq.14-2 and Eq.14-4, the heat quantities of the cooling and heating processes are absolutely equal.

式（14-2）と式（14-4）より，冷却と加熱の変化過程での熱量は完全に等しくなる．

$$q_{4-1} = -q_{2-3} \tag{14-5}$$

Furthermore, the entropy changes during the constant volume processes are the same.

さらに，それぞれの等容変化過程においてエントロピー変化も等しくなる．

$$(s_3 - s_2) + (s_1 - s_4) = c_v \ln \frac{T_b}{T_a} + c_v \ln \frac{T_a}{T_b} = 0 \quad \Rightarrow \quad s_2 - s_1 = s_3 - s_4 \tag{14-6}$$

Since there are the same temperature level processes in the cooling and heating operation for both constant volume changes, under the condition of quasi-static operation, we can reuse the negative quantity of heat due to cooling as a resource for heating. This means that heat quantities during the heating and cooling processes have no effect on the thermal efficiency expressed in Eq.14-7.

二つの定容変化の冷却と加熱の過程中には同じ温度レベルでの操作が存在しているので，準静的な熱力学的操作を前提とすれば，冷却による負の熱量を加熱のための熱源として再活用することが可能である．このことは，加熱と冷却の変化過程での熱量は，式（14-7）に示してあるスターリングサイクルの熱効率に関与しないことを示している．

$$\eta_{Stirling} = \frac{W_{Stirling}}{q_a} = \frac{|q_a| - |q_b|}{|q_a|} = 1 - \frac{T_b (s_3 - s_4)}{T_a (s_2 - s_1)} = 1 - \frac{T_b}{T_a} \tag{14-7}$$

Stirling cycle is a kind of reversible cycle whose thermal efficiency is equivalent to a Carnot cycle operated under the same temperature condition.

スターリングサイクルは，可逆サイクルの一つであるが，さらにその熱効率は，同じ温度条件下で運用されるカルノーサイクルと等しくなっている．

14.2 Internal Combustion Engine

14.2 内燃機関

Cycle analysis of internal combustion engine is one of the engineering targets for the development of classical thermodynamics. A typical internal combustion engine cycle consists of four

内燃機関のサイクル解析は，古典熱力学を確立させることに対する一つの工学的な目標である．典型的な内燃機関のサイクルは，ピストンの四つの動きで構成されている．それ

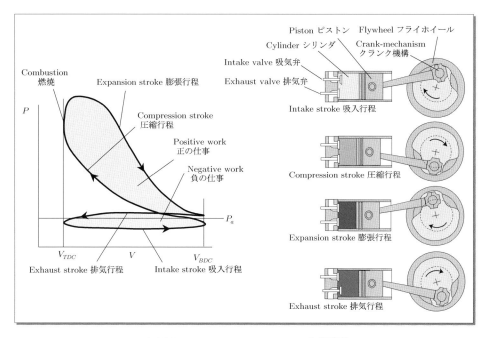

Fig.14-2 Internal combustion engine　内燃機関

piston movement strokes: intake stroke, compression stroke, expansion stroke and exhaust stroke. The piston movement and P-v chart of this engine cycle is shown in Fig.14-2.

Piston movement coupled with crank mechanism is connected to the rotation motion of the flywheel. In a four-stroke engine, two revolutions of the flywheel are divided into four piston movements. The minimum volume of working gas is given by the piston position at top dead center (TDC), and the maximum volume is given at a bottom dead center (BDC) during flywheel rotation. The compression and expansion strokes constitute a thermodynamic forward cycle to produce positive work, while the exhaust and intake strokes constitute a reverse cycle to receive negative work.

The energy flow in a four-stroke internal combustion engine cycle is given in Fig.14-3.

The fuel's heat of combustion is supplied to the working gas and it is converted to work dur-

らは，吸気行程，圧縮行程，膨張行程，排気行程の四つである．この機関サイクルのピストンの動きと P-v 線図が，図 14-2 に示してある．

ピストンの動きはクランク機構を介してフライホイール（はずみ車）の回転運動と連動している．フライホイールの 2 回転がピストンの四つの運動によって分けられ，このタイプの機関を 4 行程機関とよんでいる．作動気体の最小容積は，フライホイールの回転中での上死点（TDC）のピストン位置で与えられ，最大容積は，下死点（BDC）で与えられている．圧縮行程と膨張行程は，正の仕事を生みだすための熱力学的な順方向サイクルを形成し，排気と吸気は，負の仕事を受け取る逆サイクルを構成している．

4 行程内燃機関サイクルでのエネルギーの流れを図 14-3 に示す．燃料の燃焼熱は，作動気体に供給され，それは膨張行程の際に仕事に変換される．吸入，圧縮，排気の各行程

ing the expansion stroke. As for the intake, compression and exhaust strokes, the piston movement is caused by rotation energy of the flywheel.

では，フライホイールの回転エネルギーによってピストンの動きが引き起こされている．

Since the piston movement is dependent on the mechanical movement of the crank, the compression ratio defined by Eq.14-8 is a typical parameter of the real cycle.

ピストンの動きはクランクの機械的な動きにより決められているので，式（14-8）に示した圧縮比が実際のサイクルの特徴を表すパラメータである．

$$\text{Compression ratio} \quad 圧縮比 \quad \varepsilon = \frac{V_{BDC}}{V_{TDC}} \tag{14-8}$$

An ideal gas of G [kg] is usually considered as the working gas of an engine cycle. Then, Eq.14-9 is used for the following analysis.

G [kg] の理想気体をエンジンサイクルの作動気体として想定しているので，その状態方程式により以下の解析を行う．

$$PV = GRT \tag{14-9}$$

The effective work that is obtained through all four strokes of the cycle is the actual output work from the cycle and it is obtained from Eq.14-10.

サイクルの四つの行程の全体より得られる有効仕事が，4行程機関から得られる実質的な出力となり，それは式（14-10）より得られる．

Effective work   有効仕事

$$W_e^G = GW_e = \left[\text{Positive work} \quad 正の仕事\right] - \left[\text{Negative work} \quad 負の仕事\right] \tag{14-10}$$

Moreover, the mean effective pressure derived by Eq.14-11 is usually used as a performance parameter of the cycle.

さらに，式（14-11）より得られる平均有効圧が，このサイクルの性能を表すパラメータとして用いられている．

Mean effective pressure   平均有効圧

$$P_m = \frac{W_e^G}{V_{BDC} - V_{TDC}} \tag{14-11}$$

Since the wall temperature of the cylinder is

シリンダの壁面温度の制御は行われず，通

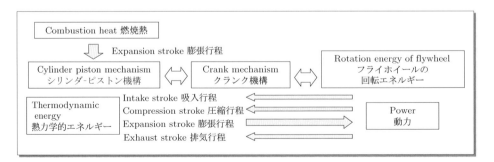

Fig.14-3   Energy flow in an internal combustion engine   内燃機関におけるエネルギーの流れ

not controlled and is usually lower than that of the working gas, adiabatic compression is difficult to achieve in a real engine cycle. Thus, we assume a polytropic process during the compression stroke. From the definition of polytropic change, the exponent "n" is obtained by Eq.14-14.

常は作動気体の温度より低いので，実際の機関サイクルでは断熱圧縮を行うことが困難である．そこで，ポリトロープ変化過程を圧縮行程に想定する．ポリトロープ変化の定義より，その指数 n は式（14-14）より得られる．

$$PV^n = \text{const.} \tag{14-12}$$

$$\ln P + n \ln V = C \tag{14-13}$$

$$n = \frac{\ln P_2 - \ln P_1}{\ln V_2 - \ln V_1} \tag{14-14}$$

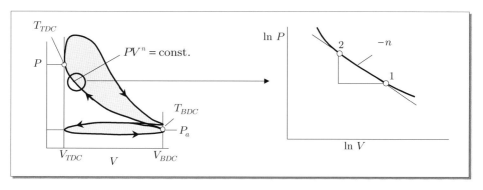

Fig.14-4 Polytropic exponent for compression stroke　圧縮行程におけるポリトロープ指数

As shown in Fig.14-4, the polytropic exponent "n" can be obtained as a negative gradient of pressure in the $\ln P$-$\ln V$ chart. Using the experimentally obtained value of "n" from a real engine cycle, the gas temperature at the end of compression can be obtained by Eq.14-15.

図 14-4 に示したとおり，ポリトロープ指数 n は P-V 対数線図の圧力の負の傾きとして得ることができる．実際の機関サイクルから実験的に得られた指数 n を用いることで，圧縮終わりの気体温度は，式（14-15）から求めることが可能になる．

$$\left.\frac{T_{TDC}}{T_{BDC}}\right|_{intake} = \left(\frac{V_{BDC}}{V_{TDC}}\right)^{n-1} = \varepsilon^{n-1} \tag{14-15}$$

14.3 Constant Volume Cycle

14.3 定容サイクル

It is known that the thermodynamic cycle of spark ignition engine cycle or gasoline engine cycle can be simulated by a constant volume (isochoric) cycle. This cycle is also called the Otto cycle honoring the inventor of the gasoline engine. The P-v and T-s charts of a constant

火花点火機関のサイクルまたはガソリン機関のサイクルは，定容（等容）サイクルで近似できることが知られている．このサイクルは，ガソリン機関を最初に開発した人の功績をたたえてオットーサイクルともよばれている．質量 1 kg の理想気体を作動媒体とする等容

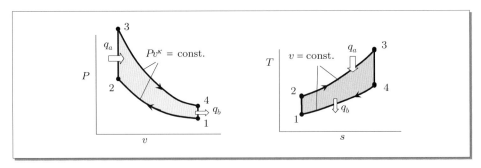

Fig.14-5   Constant volume cycle   定容サイクル

volume cycle for a working media of 1 kg ideal gas are indicated in Fig.14-5.

In a spark ignition engine, spontaneous combustion follwing the ignition takes place in the working gas of minimum volume at TDC. Thus, it can be simulated with a constant volume heating process shown by Eq.14-6.

サイクルの P-v 線図と T-s 線図が図 14-5 に示されている.

火花点火機関では,着火に続く速やかな燃焼が TDC における最小容積の作動気体の中で発生する.したがって,これは式(14-16)に示した定容の加熱変化過程で近似することができる.

$$\text{Heat supplied} \quad q_a = q_{2\text{-}3} = c_v\left(T_3 - T_2\right) \tag{14-16}$$

Heat removal from the cycle is done during the exhaust and intake processes, and this can be simulated with a constant volume cooling process shown by Eq.14-7.

機関からの熱の除去は,排気および吸気の両者の変化過程で行われ,これは式(14-17)に示した等容の冷却変化過程で近似することが可能である.

$$\text{Heat removed} \quad q_b = q_{4\text{-}1} = c_v\left(T_1 - T_4\right) \tag{14-17}$$

Using Eq.14-18 and Eq.14-19, the thermal efficiency is given by Eq.14-20.

式(14-18)と式(14-19)を用いることで,このサイクルの熱効率を式(14-20)で与えることができる.

$$\text{Effective work} \quad W_e = |q_a| - |q_b| \tag{14-18}$$

$$\text{Compression ratio} \quad \varepsilon_{Otto} = \frac{v_1}{v_2} = \frac{v_4}{v_3} = \left(\frac{T_2}{T_1}\right)^{\frac{1}{\kappa-1}} = \left(\frac{T_3}{T_4}\right)^{\frac{1}{\kappa-1}} \tag{14-19}$$

Theoretical thermal efficiency

$$\eta_{Otto} = \frac{W_e}{q_a} = 1 - \frac{T_4 - T_1}{T_3 - T_2} = 1 - \frac{T_1}{T_2} \cdot \frac{\dfrac{T_4}{T_1} - 1}{\dfrac{T_3}{T_2} - 1} = 1 - \frac{T_1}{T_2} = 1 - \frac{1}{\varepsilon_{Otto}^{\kappa-1}} \tag{14-20}$$

Both expansion and compression processes occur as adiabatic processes, and no entropy changes are expected except during the constant

膨張と圧縮の両方の変化過程とも,断熱変化過程として行われている.さらに,加熱と冷却の定容変化過程を除けば,エントロピー

volume heating and cooling processes. Thus, the entropy change for the constant volume cycle can be expressed as follows:

変化はあり得ない．したがって，定容サイクルのエントロピー変化は，つぎのように表現できる．

$$s_a = s_{2-3} = \int_2^3 \frac{dq}{T} = \int_2^3 \frac{du}{T} = c_v \int_2^3 \frac{dT}{T} = c_v \ln \frac{T_3}{T_2} \tag{14-21}$$

$$s_b = s_{4-1} = \int_4^1 \frac{dq}{T} = \int_4^1 \frac{du}{T} = c_v \int_4^1 \frac{dT}{T} = c_v \ln \frac{T_1}{T_4} \tag{14-22}$$

According to the similar method of derivation explained in Chapter 13, the following reversible cycle verification can be done for the cycle.

第 13 章で示したのと同様な手法によれば，等容サイクルが可逆サイクルであることの証明をつぎのように行うことができる．

$$s_a - s_b = 0 \quad (\text{reversible cycle}) \tag{14-23}$$

14.4 Constant Pressure Cycle

14.4 定圧サイクル

Fuel injection into a high temperature working gas (air) at the end of compression is a distinctive feature of the Diesel engine. Self-ignition of the fuel takes place owing to high temperature. Fuel injection and combustion continue during the expansion stroke. This process can be simulated with a heating process under constant pressure expansion. Since the expansion stroke with heating is a typical phenomenon of a diesel engine, this cycle is called as a diesel engine cycle. The details of the cycle are given in Fig.14-6.

ディーゼル機関は，圧縮終了時の高温作動気体（空気）中に燃料を噴射することに特徴がある．燃料の自己着火が高温により発生し，燃料の噴射とその燃焼が膨張行程の間継続して進行する．そして，この変化過程は定圧膨張の条件下での加熱変化過程で近似することができる．加熱をともなう膨張行程が，ディーゼル機関の特徴的な現象であるため，このサイクルは，ディーゼルサイクルとよばれている．サイクルの詳細は，図 14-6 に示してある．

The expansion process of the constant pressure (isobaric) cycle is divided into a constant pressure heating process and an adiabatic expansion process. Thus, the quantity of heat supplied

定圧（等圧）サイクルの膨張変化過程は，定圧の加熱過程と断熱膨張過程に分けられる．したがって，燃料の燃焼によってサイクルに供給される熱量は，式（14-24）で与えられる．

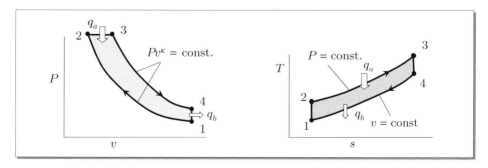

Fig.14-6　Constant pressure cycle　定圧サイクル

to the cycle from fuel combustion is given by
Eq.14-24.

$$\text{Heat supplied}\quad q_a = q_{2\text{-}3} = c_p\left(T_3 - T_2\right)\tag{14-24}$$

The heat removed, effective work and theoretical thermal efficiency can be given as follows:

サイクルから除去される熱, 有効仕事, 理論熱効率は以下のように与えられる.

$$\text{Heat removed}\quad q_b = q_{4\text{-}1} = c_v\left(T_1 - T_4\right)\tag{14-25}$$

$$\text{Effective work}\quad W_e = |q_a| - |q_b|\tag{14-26}$$

$$\text{Theoretical thermal efficiency}\quad \eta_{Diesel} = \frac{W_e}{q_a} = 1 - \frac{T_4 - T_1}{\kappa(T_3 - T_2)}\tag{14-27}$$

Using the compression ratio and the isobaric expansion ratio defined specifically for this cycle, the thermal efficiency can be rearranged to Eq.14-31.

圧縮比とこの等圧サイクルのために特に定義されている等圧膨張比を用いれば, サイクルの熱効率を式 (14-31) のように書き改めることができる.

$$\text{Compression ratio}\quad \varepsilon_{Diesel} = \frac{v_1}{v_2}\tag{14-28}$$

Isobaric expansion ratio (Cut-off ratio) 等圧膨張比 (締切り比)

$$\sigma = \frac{v_3}{v_2}\tag{14-29}$$

$$\frac{T_2}{T_1} = \varepsilon_{Diesel}^{\kappa-1},\quad \frac{T_3}{T_2} = \sigma,\quad \frac{T_4}{T_3} = \left(\frac{v_3}{v_4}\right)^{\kappa-1} = \left(\frac{v_3}{v_1}\right)^{\kappa-1} = \left(\sigma\frac{v_2}{v_4}\right)^{\kappa-1}\tag{14-30}$$

$$\eta_{Diesel} = 1 - \frac{1}{\varepsilon_{Diesel}^{\kappa-1}}\cdot\frac{\sigma^{\kappa} - 1}{\kappa(\sigma - 1)}\tag{14-31}$$

14.5 Combined Cycle

The constant volume cycle and the constant pressure cycle are two typical cycles. However, an actual engine cycle usually shows a combination of both cycles. Thus, a combined cycle, also called a Sabathe cycle, is proposed to obtain a theoretical analysis with a high level of approximation. The typical P-v and T-s charts of combined cycle are given by Fig.14-7.

It is clear that the typical internal combustion engine cycle shown in Fig.14-2 can be well simulated by the Sabathe cycle. The quantity of heat supplied is given by Eq.14-32 which is a combination of Eq.14-16 and Eq.14-24.

14.5 合成サイクル

定容サイクルと定圧サイクルは, 二つの典型的なサイクルである. しかし, 実際の機関のサイクルは, 平均的には両方のサイクルを兼ね備えたサイクル挙動を示す. したがって, サバテサイクルといわれている合成サイクルが, 近似度の高い理論的な解析を行うために提案されている. 合成サイクルの典型的な P-v 線図と T-s 線図を図 14-7 に示す.

図 14-2 に示した内燃機関の典型的なサイクルが, サバテサイクルにより良好に近似できることがあきらかである. 供給される熱量は, 式 (14-32) で与えられるが, これは, 式 (14-16) と式 (14-24) を組合せたものである.

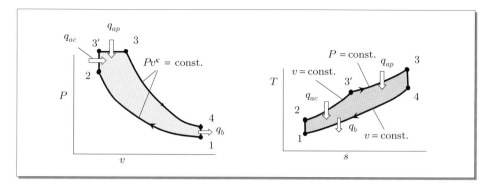

Fig.14-7 Combined cycle　合成サイクル

Heat supplied
$$q_a = q_{ac} + q_{ap} = q_{2-3'} + q_{3'-3} = c_v \left(T_{3'} - T_2 \right) + c_p \left(T_3 - T_{3'} \right) \tag{14-32}$$

Theoretical expressions of heat removed and effective work are given as follows:

除去される熱と有効仕事の理論式は，以下のように与えられている．

Heat removed　$q_b = q_{4-1} = c_v \left(T_1 - T_4 \right)$　$(14\text{-}33)$

Effective work　$W_e = |q_a| - |q_b|$　$(14\text{-}34)$

In addition to the compression ratio and the isobaric expansion ratio, the pressure rise ratio is introduced into the Sabathe cycle.

圧縮比と等圧膨張比に加えて，圧力上昇比をサバテサイクルに新たに導入する．

ξ　Pressure rise ratio 圧力上昇比
$$\varepsilon = v_1 / v_2, \quad \xi = P_{3'} / P_2, \quad \sigma = v_3 / v_{3'} \tag{14-35}$$

Then, the thermal efficiency for the Sabathe cycle can be rearranged as follows:

その結果，サバテサイクルについての熱効率は，つぎのように書き改められる．

Theoretical thermal efficiency
$$\eta_{Sabathe} = \frac{W_e}{q_a} = 1 - \frac{\left(1 / \varepsilon^{\kappa-1} \right) \left(\xi \sigma^\kappa - 1 \right)}{\left(\xi - 1 \right) + \kappa \xi (\sigma - 1)} \tag{14-36}$$

Finally, comparisons of the thermal efficiencies of different cycles can be done under various conditions. Some of the results are shown here.

最終的に種々の条件下におけるサイクル間の熱効率の比較を行うことができるようになり，そのいくつかの結果をここに示す．

$$\left[T_1, P_1, \varepsilon, q_a \right]_{Otto} = \left[T_1, P_1, \varepsilon, q_a \right]_{Diesel} = \left[T_1, P_1, \varepsilon, q_a \right]_{Sabathe}$$
$$\Rightarrow \quad \eta_{Otto} > \eta_{Sabathe} > \eta_{Diesel} \tag{14-37}$$

$$\left[T_1, P_3, q_a \right]_{Otto} = \left[T_1, P_3, q_a \right]_{Diesel} = \left[T_1, P_3, q_a \right]_{Sabathe}$$
$$\Rightarrow \quad \eta_{Diesel} > \eta_{Sabathe} > \eta_{Otto} \tag{14-38}$$

14.6 Joule Cycle

The gas turbine engine is one type of an internal combustion engines, however, it is not a reciprocating engine. As shown in Fig.14-8, it has a compressor for the compression process and a turbine for the expansion process. A turbine shaft connects the compressor with the output load. The combustor is located on between the compressor and the turbine.

A working gas flows into the compressor; its pressure and temperature are elevated due to compression. Constant pressure heating proceeds in the combustor. In the turbine, the working gas expands. Its temperature and pressure decrease with the adiabatic expansion work.

The P-v and T-s charts of this thermodynamic cycle are shown in Fig.14-9. An ideal gas of 1 kg mass is considered as the working gas of

14.6 ジュールサイクル

ガスタービン機関は，内燃機関の一種であるが往復動機関ではない．図 14-8 に示してあるとおり，ガスタービン機関は，圧縮過程のための圧縮機と膨張過程のためのタービンをもっている．タービンの回転軸は，圧縮機と出力としての負荷に繋がっている．また，圧縮機とタービンの間には，燃焼器が置かれている．

作動ガスは圧縮機に流入し，圧縮により温度と圧力が上昇する．燃焼器では等圧加熱が進行する．タービンの中では作動ガスが膨張する．タービン内では作動ガスの温度と圧力が断熱膨張仕事とともに低下する．

この熱力学的サイクルの P-v 線図と T-s 線図が図 14-9 に示してある．この場合，質量 1 kg の理想気体を作動ガスとしている．この

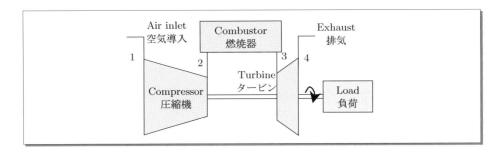

Fig.14-8 Gas turbine ガスタービン

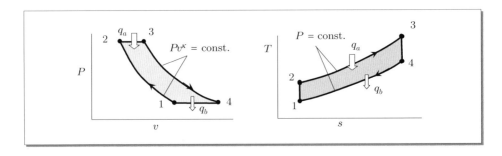

Fig.14-9 Joule cycle ジュールサイクル

the cycle. This gas turbine cycle is called Joule cycle or Brayton cycle.

In a Joule cycle, both heat addition and heat removal occur during constant pressure processes. Thus, Eq.14-39 and Eq.14-40 are the basic equations of the thermodynamic analysis.

ガスタービンサイクルは，ジュールサイクルまたはブレイトンサイクルとよばれている．

ジュールサイクルでは，熱の供給と除去の両方が等圧変化過程の間に行われる．したがって，式（14-39）と式（14-40）が熱力学的な解析の基礎的な式である．

$$\text{Heat supplied} \quad q_a = q_{2\text{-}3} = c_p(T_3 - T_2) \tag{14-39}$$

$$\text{Heat removed} \quad q_b = q_{4\text{-}1} = c_p(T_1 - T_4) \tag{14-40}$$

Since the cycle has two working processes, the effective work is given by Eq.14-43.

このサイクルは，仕事を行う二つの変化過程をもつので，有効仕事は式（14-43）によって与えられる．

Negative work to compressor　圧縮機への負の仕事
$$W_C = h_2 - h_1 = c_p(T_2 - T_1) \tag{14-41}$$

Positive work to turbine　タービンへの正の仕事
$$W_T = -(h_4 - h_3) = c_p(T_3 - T) \tag{14-42}$$

Effective work　$W_e = |W_T| - |W_C| = |q_a| - |q_b|$ (14-43)

Using the pressure ratio and the compression ratio defined for the cycle, the thermal efficiency is given by Eq.14-47. This expression is similar to the thermal efficiency of a constant volume cycle for a reciprocating engine.

このサイクルに対して定義されている圧力比と圧縮比を用いれば，熱効率は，式（14-47）で与えられる．この表現は，往復動機関の定容サイクルの熱効率とまったく同じである．

$$(\text{Pressure ratio　圧力比}) \quad \gamma = \frac{P_2}{P_1} = \frac{P_3}{P_4} \tag{14-44}$$

$$(\text{Compression ratio　圧縮比}) \quad \varepsilon = \frac{v_1}{v_2} = \frac{v_4}{v_3} \tag{14-45}$$

$$\frac{T_1}{T_2} = \left(\frac{P_1}{P_2}\right)^{\frac{\kappa-1}{\kappa}} = \left(\frac{P_4}{P_3}\right)^{\frac{\kappa-1}{\kappa}} = \frac{T_4}{T_3} = \left(\frac{1}{\lambda}\right)^{\frac{\kappa-1}{\kappa}} = \frac{1}{\varepsilon^{\kappa-1}} \tag{14-46}$$

Theoretical thermal efficiency　理論熱効率

$$\eta_{Brayton} = \frac{W_e}{q_a} = 1 - \frac{T_4 - T_1}{T_3 - T_2} = 1 - \frac{T_1}{T_2} \cdot \frac{\frac{T_4}{T_1} - 1}{\frac{T_3}{T_2} - 1} \tag{14-47}$$

$$= 1 - \frac{T_1}{T_2} = 1 - \left(\frac{1}{\lambda}\right)^{\frac{\kappa-1}{\kappa}} = 1 - \frac{1}{\varepsilon^{\kappa-1}}$$

Problems	問題

△14-1

If the compression ratio is 9.9, determine the thermal efficiency of an Otto cycle. Here, the ratio of specific heat of the working fluid is 1.4

△14-2

If the compression ratio is 18.0, and the isobaric expansion ratio is 4.0, determine the thermal efficiency of a Diesel cycle. Here, the ratio of specific heat of the working fluid is 1.4.

△14-3

If the compression ratio is 18.0, the isobaric expansion ratio is 2.0, and the pressure rise ratio is 5.0, determine the thermal efficiency of a Sabathe cycle. The ratio of specific heat of the working fluid is 1.4.

△14-4

If the pressure ratio is 12.0, determine the thermal efficiency of a Brayton cycle. The ratio of specific heat of the working fluid is 1.4

△14-1

圧縮比が 9.9 であるオットーサイクルの熱効率を求めなさい．作動流体の比熱比は 1.4 とする．

△14-2

圧縮比が 18.0 で等圧膨張比が 4.0 であるディーゼルサイクルの熱効率を求めなさい．作動流体の比熱比は 1.4 とする．

△14-3

圧縮比が 18.0，等圧膨張比が 2.0 で圧力上昇比が 5.0 であるサバテサイクルの熱効率を求めなさい．作動流体の比熱比は 1.4 とする．

△14-4

圧力比が 12.0 であるブレイトンサイクルの熱効率を求めなさい．作動流体の比熱比は 1.4 とする．

Chapter 15
Steam and Steam Cycle
蒸気と蒸気サイクル

15.1 Steam

Steam is a working media widely used in thermodynamic prime mover during the early days and in modern power plant. Steam is a gaseous state of water and its general characteristics have already been shown in the phase diagram in chapter 3 (See Fig.3-2). The phase change from liquid state (water) to gaseous state (steam or water vapor) is shown in more detail, using the illustration and the P-v chart in Fig.15-1.

Now, we consider an isobaric phase change of water. The liquid state is stable when its pressure and temperature are kept constant, and it is generally called a compressed liquid. A tem-

15.1 蒸 気

蒸気は，以前は熱力学的原理による原動機において，そして現代における発電用原動機において広く用いられている作動流体である．蒸気は水が気体状態になったものであり，蒸気の一般的な特性は，第3章の相図（図3-2参照）ですでに明らかにされている．液体状態（水）から気体状態（蒸気または水蒸気）へのさらにくわしい相変化を，図15-1に図解と P-v 線図を用いて示してある．

ここで，水が等圧で相変化するとしよう．液体状態は，その圧力と温度が一定ならば安定であり，この液体状態は，一般的には圧縮液とよばれている．液相においての温度上昇

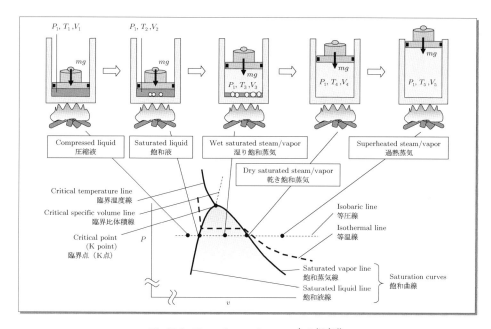

Fig.15-1 Phase change of water　水の相変化

perature increase in the liquid phase results in a slight increase in volume.

When the temperature of liquid water reaches the boiling temperature due to heating, a gaseous phase such as bubbles of water vapor or steam appears in the liquid phase. With more heat added, more bubbles are produced and no more temperature increase is observed. In other words, this is so called constant pressure boiling and Gibbs' free energy is kept constant ($dg = 0$ at Eq.6-46). During this heat addition, all of the supplied heat is consumed to enhance the phase change from liquid to gaseous state. The water vapor or steam that co-existed with boiling water is a saturated vapor. Wet saturated vapor means a mixture of saturated water and saturated vapor.

After all the water is vaporized, the temperature of the steam starts to increase with heating. Steam that has a temperature higher than the boiling temperature is called superheated steam.

Both saturated liquid and saturated vapor lines in Fig.15-1 characterize the phase change from liquid to gaseous state. As the operating pressure increases, the co-existing zone of wet saturated vapor becomes narrow. At critical pressure, this zone vanishes.

15.2 Equation of State for Real Gas

Isothermal lines near the wet saturated vapor zone are illustrated in a P-v chart shown in

は，体積のわずかな増加をもたらす．

加熱によって液体としての水の温度が沸点に達すると，水蒸気の気泡のような気相が液相のなかに出現する．さらに熱を加えると，気泡は増加するが温度上昇は見られなくなる．言い換えると，これは定圧沸騰とよばれている過程であり，ギブスの自由エネルギーは一定に保たれている（式(6-46)にて $dg = 0$）．この加熱の状況では，供給された熱のすべてが液体から気体状態への相変化を促進するために費やされている．沸騰している水と共存している水蒸気または蒸気は，飽和蒸気である．一方，湿り飽和蒸気は飽和液と飽和蒸気の混合物のことである．

すべての水が蒸発した後，加熱とともに蒸気の温度は上昇を開始する．沸点の温度より高い温度となる蒸気は，過熱蒸気とよばれている．

図 15-1 に示してある飽和液線と飽和蒸気線の両者が，液体から気体状態への相変化を特徴づけている．現象を操作している場の圧力が増加するにしたがい，湿り飽和蒸気の共存領域はせまくなり，臨界圧力のところでこの領域は消滅する．

15.2 実存気体の状態方程式

湿り飽和蒸気の領域付近の等温線を図 15-2 に示した P-v 線図のなかに書き込んで示す．

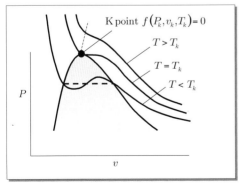

Fig.15-2 Isothermal lines of water vapor
水蒸気の等温線

Fig.15-2. They show different behaviors corresponding to the level of temperature.

Furthermore, it is clear that these lines can not be expressed with hyperbolic isothermal lines given by the equation of state of an ideal gas. As explained in chapter 3 (See Fig.3.1), the molecular volume of a real gas such as steam can not be ignored when the gas has a thermodynamic condition near the saturated vapor line. Then, we need another formula for Eq.15-1 that can explain the relationships among pressure, specific volume and temperature of a real gas.

等温線は，その温度レベルより異なる挙動を示している．

さらに，理想気体の状態方程式により与えられる双曲線の等温線では，この等温線を表すことができないことが明らかである．第3章で述べたように（図 3.1 参照），実存気体の熱力学的条件が飽和蒸気線に近いものである場合は，蒸気のような実存気体では分子の体積を無視することができない．したがって，実存気体の体積，比容積，温度の間の関係を与える式（15-1）に対する別の具体的な関係式が必要になる．

$$f(P, v, T) = 0 \tag{15-1}$$

One of the formulas of the equation of state for a real gas is given as van der Waals' equation of state. It is given as follows:

実存気体の状態方程式の一つにファン・デル・ワールスの状態方程式がある．この式は，つぎのように与えられている．

van der Waals' equation of state
ファン・デル・ワールスの状態方程式

$$\left(P + \frac{a}{v^2}\right)(v - b) = RT \tag{15-2}$$

$$f(P, v, T) = v^3 - \left(\frac{RT}{P} + b\right)v^2 + \frac{a}{P}v - \frac{ab}{P} = 0 \tag{15-3}$$

It is clear that van der Waals' equation of state is developed from the equation of state of an ideal gas. The pressure of an ideal gas is the pressure caused by the average momentum change of molecular collisions. However, the measured pressure of a real gas is the wall pressure caused by molecular collisions and the negative van der Waals' force between the wall and the molecules. Thus, the pressure of a real gas is corrected by a parameter "a". The volume of an ideal gas is the space where molecules move. Since this space is reduced with the volume of the molecules themselves, the actual space for molecular movement of a real gas is corrected by a parameter "b".

The van der Waals' equation of state can explain the various isothermal lines illustrated in Fig.15-2. The isothermal line that is passing

ファン・デル・ワールスの状態方程式は，理想気体の状態方程式より発展したものであることが明らかである．理想気体の圧力は，分子の衝突による平均的な運動量変化に起因する圧力である．しかし，実存気体の計測される圧力は壁面の圧力であり，それは分子の衝突と，分子と壁の間に働く負のファン・デル・ワールス力に起因している．したがって，実存気体の圧力が修正パラメータ a で修正されている．理想気体の体積は，分子の動く空間を意味している．この空間は，分子自身の体積により減少するので，実存気体の分子運動のための実際の空間は，パラメータ b により修正されている．

ファン・デル・ワールスの状態方程式は，図 15-2 に記してある種々の等温線を説明することができる．臨界点（P_k, v_k, T_k）を通る

through the critical point (P_k, v_k, T_k) has to satisfy the following conditions:

等温線は，つぎのような条件を満さなければならない．

$$\left(\frac{\partial P}{\partial v}\right)_T\Big|_{v=v_k} = 0, \quad \left(\frac{\partial^2 P}{\partial v^2}\right)_T\Big|_{v=v_k} = 0 \tag{15-4}$$

From these conditions, we can obtain the relationship among the thermodynamic quantities of state and the parameters of the van der Waals' equation of state.

このような条件により，ファン・デル・ワールスの状態方程式中のパラメータと，熱力学的状態量の関係を求めることが可能になる．

$$a = 3P_k v_k^2, \quad b = \frac{v_k}{3}$$
$$\Rightarrow \quad 4P_k \cdot \frac{2}{3} v_k = RT_k, \quad R = \frac{8}{3} \cdot \frac{P_k v_k}{T_k} \tag{15-5}$$

The value of "a" means that the pressure caused by the average momentum change of molecular collisions is, at the critical point, 4 times larger than the measured critical pressure P_k. Also the value of "b" means that the total volume of the molecules themselves is one third of the thermodynamic volume of gaseous state at critical point (See the moleculer dynamics and pressure explained by Fig.5-2).

パラメータ a の値は，分子衝突による平均的な運動量変化に起因する圧力が，臨界点において，実際に計測される臨界圧力 P_k の4倍であることを意味している．また，パラメータ b の値は，分子自身の総体積が臨界点における気体状態の熱力学的体積の3分の1であることを意味している（図5-2で説明してある分子動力学と圧力を参照のこと）．

There are many kinds of expressions of the equation of state for a real gas. The following are two of the typical equations. The equation of state by Berthelot is similar to the van der Waals' equation of state but the temperature effect on pressure is introduced. The equation of state by Virial expansion form can express a more accurate thermodynamic state but its theoretical background is not clear.

実存気体の状態方程式には，種々の表現形式がある．以下に述べる二つは，そのなかの典型的なものである．ベルトゥローの状態方程式は，ファン・デル・ワールスの状態方程式に近いが，圧力に対する温度影響が加えられている．ビリアルの展開式による状態方程式は，熱力学的状態を精度よく表現できるが，その理論的な根拠は不明瞭である．

Equation of state by Berthelot form ベルトゥローによる状態方程式

$$\left(P + \frac{a'}{Tv^2}\right)(v - b) = RT \tag{15-6}$$

Equation of state by Virial expansion form ビリアル展開式による状態方程式

$$\frac{Pv}{RT} = 1 + \frac{B(T)}{v} + \frac{C(T)}{v} + \frac{D(T)}{v} + \cdots \tag{15-7}$$

15.3 Wet Saturated Vapor

A substance of 1 kg mass that shows an apparent specific volume whose value falls between the saturated liquid and saturated vapor is a mixed substance of liquid and gas and so called "Wet Saturated Vapor/Steam". It has no original thermodynamic quantities corresponding to the given specific volume. To express the thermodynamic quantities in this situation, we need to introduce a parameter "x" that is called the quality of dryness. The definition and meaning is given by Eq.15-8.

15.3 湿り蒸気

飽和液と飽和蒸気の中間の値となるみかけの比容積をもつ質量 1 kg の物質は，液体と気体の混合物であり，これは湿り蒸気とよばれている．これは与えられた比容積に対応した固有の熱力学的状態量をもっているわけではない．この状況における熱力学的状態量を表現するために，乾き度 "x" というパラメータを導入するが，このパラメータの定義と意味は，式 (15-8) に示されている．

$$x = \frac{\text{mass of vapor}}{\text{total mass of wet saturated vapor}} = \frac{\text{飽和蒸気の質量}}{\text{湿り蒸気の全質量}}$$
$$= \frac{\text{mass of saturated vapor}}{\text{mass of saturated liquid} + \text{mass of saturated vapor}}$$
$$= \frac{\text{飽和蒸気の質量}}{\text{飽和液の質量} + \text{飽和蒸気の質量}} \tag{15-8}$$

Also, parameter "$1-x$" is co-defined and is called the quality of wetness. Then, the apparent specific volume of the mixture is given by Eq.15-9.

また，湿り度とよばれる "$1-x$" も同時に定義する．このようにすると，問題としている混合物の見かけの比容積は，式 (15-9) で与えられる．

$$v_{wet} = (1-x)v' + xv'' = v' + x(v'' - v') \tag{15-9}$$

Furthermore, an important thermodynamic property of substance to describe the thermodynamic state of a mixture is the latent heat of vaporization. When the thermodynamic quantities $(u',h',s',v',...)$ of the saturated water are given by Eq.15-10, and the thermodynamic quantities $(u'',h'',s'',V'',...)$ of the saturated gas are known, latent heat is specified by Eq.15-11,

さらに，この混合物の熱力学的状態量を記述するための物質の熱力学的物性値で重要なものに蒸発潜熱がある．飽和液の熱力学的状態量 $(u',h',s',v',...)$ が式 (15-10) で与えられ，飽和蒸気の熱力学的状態量 $(u'',h'',s'',V'',...)$ が知られているならば，蒸発潜熱は式 (15-11) により定められる．

$$u' = \int_{T_0}^{T_s} c_{vl} dT, \quad h' = \int_{T_0}^{T_s} c_{pl} dT, \quad s' = \int_{T_0}^{T_s} \frac{c_{pl}}{T} dT \tag{15-10}$$
$$r = h'' - h' = u'' - u' + P(v'' - v') \tag{15-11}$$

Using the above equations and the illustrations shown in Fig.15-3, the thermodynamic quantities of wet saturated vapor can be obtained as the thermodynamic quantities of the mixture.

上記の式や図 15-3 の説明をもとに，湿り飽和蒸気についての熱力学的状態量は，混合物の熱力学的状態量として求めることができる．さらに過熱蒸気の状態量については，式

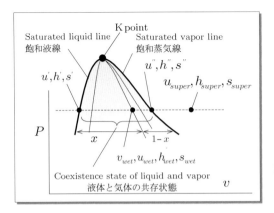

Fig.15-3 Wet saturated vapor and quality of dryness
湿り飽和蒸気と乾き度

Furthermore, the thermodynamic quantities of super heated vapor can be obtained by Eq.15-15.

（15-15）より求めることができる.

$$u_{wet} = (1-x)u' + xu'' = u' + x(u'' - u') \tag{15-12}$$

$$h_{wet} = (1-x)h' + xh'' = h' + x(h'' - h') = h' + xr \tag{15-13}$$

$$s_{wet} = (1-x)s' + xs'' = s' + x(s'' - s') = s' + x\frac{r}{T_s} \tag{15-14}$$

$$u_{super} = u'' + \int_{T_0}^{T_s} c_v dT, \quad h_{super} = h'' + \int_{T_0}^{T_s} c_p dT, \quad s_{super} = s'' + \int_{T_0}^{T_s} \frac{c_p}{T} dT \tag{15-15}$$

15.4 Thermodynamic State Change of Water Vapor

15.4 水蒸気の熱力学的状態変化

Here, consider four typical thermodynamic state changes of a real gas. Even if the equations of state for the substance corresponding to various phases are given separately, reversible change is possible. In other words, a quantity of heat supplied to the substance can be expressed by Eq.15-17 under the conditions of Eq.15-16.

ここで，実存気体の四つの典型的な熱力学的状態変化を検討する．物質の液相や気相などの各種の相に対応して，個別に状態方程式が与えられているとしても，可逆変化は可能である．言い換えれば，式（15-16）の条件下において，物質に供給される熱量は，式（15-17）で表すことができる.

$$f_{liquid}(P,v,T) = 0, \quad f_{wet\ saturated\ vapor}(P,v,T) = 0, \quad f_{super\ heated\ vapor}(P,v,T) = 0 \tag{15-16}$$

$$dq = Tds = dh - vdP = du + Pdv \tag{15-17}$$

The P-v and T-s charts of isothermal change are shown in Fig.15-4. As mentioned in the previous chapters, the characteristic equation of isothermal change is given by Eq.15-18. The quantities of heat, expansion work and technical work are given as follows:

等温変化の P-v 線図と T-s 線図を図 15-4 に示す．これまでの章で述べたように，等温変化の特性方程式は式（15-18）で与えられていて，熱量，膨張仕事，工業仕事は，また以下のように与えられている.

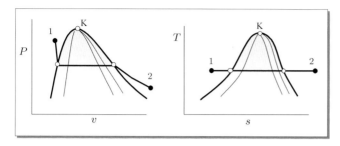

Fig.15-4 Isothermal change
等温変化

$$dT = 0 \tag{15-18}$$

$$q_{1-2} = \int_1^2 Tds = T(s_2 - s_1) \tag{15-19}$$

$$\begin{aligned} W_{1-2} &= \int_1^2 Pdv = \int_1^2 \{dq - dh + d(pv)\} \\ &= q_{1-2} - (h_2 - h_1) + (P_2 v_2 - P_1 v_1) \\ &= q_{1-2} - (u_2 - u_1) \end{aligned} \tag{15-20}$$

$$L_{1-2} = -\int_1^2 vdP = q_{1-2} - (h_2 - h_1) \tag{15-21}$$

As for an isobaric change, which is a constant pressure change as shown by Fig.15-5, the thermodynamic quantities are given by the following equations:

等圧変化については，これは図 15-5 で示されている圧力一定の変化であるが，熱力学的諸量については，以下の式によって与えられる．

$$dP = 0 \tag{15-22}$$

$$q_{1-2} = h_2 - h_1 \tag{15-23}$$

$$\begin{aligned} W_{1-2} &= \int_1^2 Pdv = \int_1^2 \{dq - dh + d(pv)\} \\ &= q_{1-2} - (h_2 - h_1) + P(v_2 - v_1) \\ &= P(v_2 - v_1) \end{aligned} \tag{15-24}$$

$$L_{1-2} = 0 \tag{15-25}$$

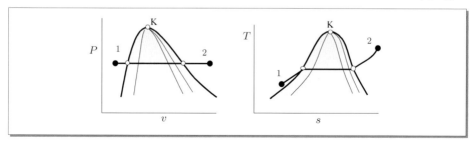

Fig.15-5　Isobaric (constant pressure) change　等圧(定圧)変化

The P-v and T-s charts of constant volume change given in Fig.15-6 show some different behaviors from previous cases. It is clear that

図 15-6 に示した等容変化の P-v 線図と T-s 線図は，いままでの場合とは若干異なった挙動を示す．乾き度が，この挙動の支配要

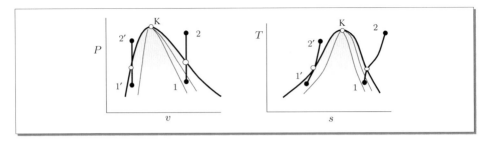

Fig.15-6 Isochoric (constant volume) change　等容(定容)変化

the quality of dryness is a controlling factor of these behaviors. When its value is small and the specific volume of wet saturated vapor is less that that of v_k, the wet saturated vapor changes to liquid phase as the pressure rises. On the other hand, the change from wet saturated vapor to superheated vapor appears with an increase of pressure when a specific volume is larger than v_k.

However, there are no differences in the equations for thermodynamic quantities. It is clear that Eq.15-26 is the characteristic equation of a constant volume change. The thermodynamic quantity changes for a change from wet saturated vapor can be derived as follows:

因であることが明らかである．乾き度の値が小さく，湿り飽和蒸気の比容積が v_k より小さい場合では，湿り飽和蒸気は圧力の上昇とともに液相へと変化する．一方，比容積が v_k より大きい場合では，圧力の増加に従い，湿り飽和蒸気から過熱蒸気への変化が現れる．

ところが，熱力学的諸量の式においては，このような変化はみられない．式 (15-26) は，等圧変化の特性方程式であり，湿り飽和蒸気からの変化に対する熱力学的諸量の変化が，以下のように求められることは明らかである．

$$dv = 0 \tag{15-26}$$

$$q_{1-2} = \int_1^2 Tds = u_2 - u_1$$
$$= (h_2 - h_1) - v(P_2 - P_1) \tag{15-27}$$

$$W_{1-2} = 0 \tag{15-28}$$

$$L_{1-2} = -\int_1^2 vdP = v(P_2 - P_1) \tag{15-29}$$

As for an adiabatic change of wet saturated vapor, the following equations are obtained and almost similar behaviors with Fig.15-6 are observed in Fig.15-7. In this case, behaviors are divided by s_K.

湿り飽和蒸気の断熱変化に関しても，以下の式が得られ，かつ図 15-7 において，図 15-6 とほぼ同様な挙動がみられる．この場合では，s_K が挙動を分割している．

$$ds = 0 \tag{15-30}$$

$$q_{1-2} = 0 \tag{15-31}$$

$$W_{1-2} = \int_1^2 Pdv = -\int_1^2 du = -(u_2 - u_1) \tag{15-32}$$

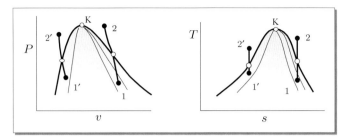

Fig.15-7 Adiabatic Change
　　　　　断熱変化

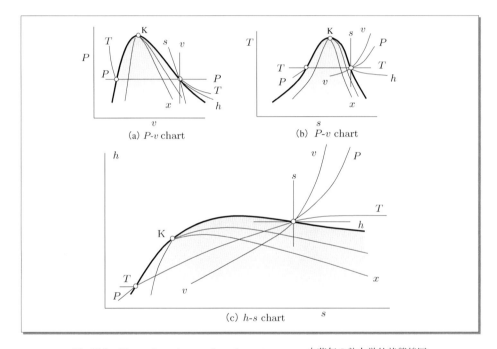

(a) P-v chart

(b) P-v chart

(c) h-s chart

Fig.15-8 Thermodynamic state charts for water vapor　水蒸気の熱力学的状態線図

$$L_{1-2} = -\int_1^2 vdP = -\int_1^2 dh = -(h_2 - h_1) \tag{15-33}$$

Negative sign in Eqs.15-32 and 15-33 correspond to an adiabatic cowpression processes shown by Fig.15-7.

Thermodynamic state charts for water vapor are summarized as shown in Fig.15-8. An enthalpy-entropy chart shown by (c) is often used to analyze the thermal treatment of steam.

式（15-32）と式（15-33）における負の符号は，図 15-7 に示してある断熱圧縮過程に対応している．

水蒸気の熱力学的状態線図は，図 15-8 のようにまとめられている．図（c）に示したエンタルピー - エントロピー線図は，蒸気の熱的な扱いを解析する場合にしばしば用いられている．

15.5 Rankine Cycle

A typical steam turbine engine is shown in Fig.15-9. The thermodynamic cycle for this thermal engine is known as the Rankine cycle. Here, consider the thermal efficiency of the cycle using the T-s and h-s charts given by Fig 15-10.

The steam turbine engine consists of a boiler, turbine, condenser and a pump. The boiler gives a quantity of heat to the working media of liquid state and produces a superheated vapor. In the turbine, adiabatic expansion of the gaseous (superheated vapor) working media produces expansion work. Passing through a condenser, the working media releases a quantity of heat and changes from the gaseous state to a liquid state. The working media in a liquid state is supplied to the boiler after its pressure is raised by the

15.5 ランキンサイクル

典型的な蒸気タービン機関を図 15-9 に示す. この熱機関に対する熱力学的サイクルは, ランキンサイクルとして知られている. ここでは, 図 15-10 で与えられている T-s 線図と h-s 線図を用いてランキンサイクルの熱効率を検討する.

蒸気タービン機関は, ボイラ, タービン, 凝縮器 (復水器), ポンプから成り立っている. ボイラは, 液体状態の作動媒体に熱量を与え, 過熱蒸気を生成する. タービンでは, 気体状態 (過熱蒸気) の作動媒体の断熱膨張により膨張仕事が行われる. 凝縮器を通過する際に, 作動媒体は熱量を放出し, 気体状態から液体状態へ変化する. ポンプにより昇圧された後, 液体状態の作動媒体はボイラに供給される. 作動媒体の循環している流れを検討すれば, 位置 4 で示される場所にて圧力が最大となる

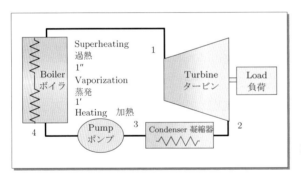

Fig.15-9 Steam turbine engine
蒸気タービン機関

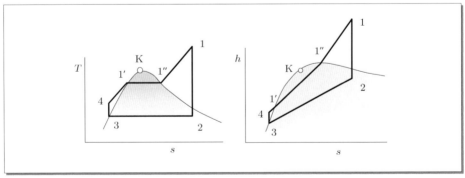

Fig.15-10 Rankine cycle   ランキンサイクル

pump. Considering a re-circulating flow of working media, it is clear that the highest pressure is observed to be at point 4.

From the enthalpy analysis of working media (in this case: water) of 1 kg mass, expansion work which is produced as turbine work is expressed by Eq.15-34.

ことが理論的に明らかである.

質量 1 kg の作動媒体（この場合は水）のエンタルピー解析より, 式 (15-34) で示されるタービン仕事として膨張仕事がなされる.

$$1 \Rightarrow 2 \left[\text{Turbine work}\right] \quad W_T = (h_2 - h_1) = -\int_1^2 v dP \tag{15-34}$$

The quantity of heat removed from the working media passing through the condenser is given by Eq.15-35.

凝縮器を通過する作動媒体から除去される熱量は, 式 (15-35) により与えられる.

$$2 \Rightarrow 3 \left[\text{Heat removed}\right] \quad q_{2\text{-}3} = h_3 - h_2 \tag{15-35}$$

The negative work needed to elevate the pressure of the working media is given by Eq.15-36.

作動媒体の圧力を上昇させるために必要な負の仕事は, 式 (15-36) により与えられる.

$$3 \Rightarrow 4 \left[\text{Pump work}\right] \quad W_C = h_4 - h_3 = \int_3^4 v dP$$
$$\cong v_3 (P_4 - P_3) \cong v_3 (P_1 - P_2) \tag{15-36}$$

During the heat receiving process of the working media in the boiler, the state of the working media is initially changed from a compressed liquid at point 4 to a saturated liquid at point 1′. After passing through a wet saturated vapor state, it changes from saturated vapor at point 1″ to superheated vapor at point 1. Then, the quantity of heat supplied is given by a summation of the quantities of heats for all the changes in the boiler and it can be expressed by Eq.15-37,

ボイラ内の作動媒体の熱を受け取る過程においては, 作動媒体の状態は位置 4 に対応した圧縮液から位置 1′に対応した飽和液にまず変化する. 湿り飽和蒸気の状態を通過した後, 作動媒体は, 位置 1″に対応した飽和蒸気状態から位置 1 に対応した過熱蒸気へと変化する. したがって, 供給した熱量はボイラ内でのすべての変化に対応した熱の和として与えられ, 式 (15-37) により表すことができる.

$$4 \Rightarrow 1 \left[\text{Heat supplied}\right] \quad q_{4\text{-}1} = (h_{1'} - h_4) + (h_{1''} - h_{1'}) + (h_1 - h_{1''})$$
$$= (h_{1'} - h_4) + r + (h_1 - h_{1''}) = h_1 - h_4 \tag{15-37}$$

The net output work from the Rankine cycle is given by Eq.15-38, and finally, the efficiency shown by Eq.15-39 is given as the thermal efficiency for the Rankine cycle.

ランキンサイクルの実質的な出力仕事は, 式 (15-38) で与えられ, 最終的に, 式 (15-39) で示される効率がランキンサイクルの熱効率として与えられる.

$$W_{Rankine} = |W_T| - |W_C| = (h_1 - h_2) - (h_4 - h_3) \tag{15-38}$$

$$\eta_{Rankine} = \frac{W_{Rankine}}{q_{4\text{-}1}} = \frac{(h_1 - h_2) - (h_4 - h_3)}{(h_1 - h_4)} = \frac{(h_1 - h_2) - (h_4 - h_3)}{(h_1 - h_3) - (h_4 - h_3)} \tag{15-39}$$

Problems

問題

△15-1

Estimate the pressure of air at 180 K and $v = 0.00285$ m³/kg on the basis of (1) the ideal-gas equation of state, and (2) the van der Waals' equation of state. Here the constants of a and b in the van der Waals' equation of state are 0.162 m⁶kPa/kg and 0.00126 m³/kg respectively, and the gas constant of air is 0.2870 kJ/ (kg·K)

△15-2

If the quality of dryness of the wet saturated vapor at 0.2 MPa is 0.5, determine the specific volume and the specific entropy. Here, values of specific volume and the specific entropy of saturated liquid and vapor at 0.2 MPa are as follows.

$$v' = 0.001061 \, \text{m}^3/\text{kg}, \quad v'' = 0.8857 \, \text{m}^3/\text{kg}$$
$$s' = 1530 \, \text{J}/(\text{kg} \cdot \text{K}), \quad s'' = 7127 \, \text{J}/(\text{kg} \cdot \text{K})$$

△15-3

Consider a Rankine cycle shown in Fig.15-11. If $h_1 = 3365$ kJ/kg, $h_2 = 2012$ kJ/kg, $h_3 \cong h_4 = 138$ kJ/kg in Fig.15-11, determine the thermal efficiency of the cycle.

△15-1

180 K で比体積 $v = 0.00285$ m³/kg の空気の圧力を, (1) 理想気体の状態式, (2) ファン・デル・ワールスの状態式を用いて推算しなさい. ここで, ファン・デル・ワールスの状態式中の定数 a と b はそれぞれ 0.162 m⁶ kPa/kg と 0.00126 m³/kg であり, 空気のガス定数は 0.2870 kJ/(kg·K) であるとする.

△15-2

0.2 MPa の湿り飽和蒸気の乾き度が 0.5 であるとき, その比体積と比エントロピーを求めなさい. ここで, 0.2 MPa における飽和水と飽和蒸気の比体積および比エントロピーは以下のとおりである.

△15-3

図 15-11 のようなランキンサイクルを考える. 図において $h_1 = 3365$ kJ/kg, $h_2 = 2012$ kJ/kg, $h_3 \cong h_4 = 138$ kJ/kg であるとき, このサイクルの熱効率を求めなさい.

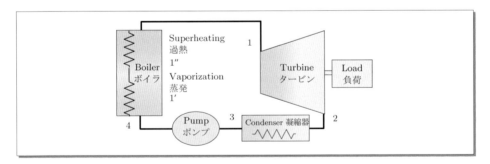

Fig.15-11 Rankine cycle for Problem 15-3
問題 15-3 のランキンサイクル

Solutions
解　答

△1-1

△1-1

(1)

（1）

$$25 + 273.15 = 298.15\,\mathrm{K}$$

(2)

（2）

$$\frac{5}{9}(86 - 32) = 30°\mathrm{C}$$

△1-2

△1-2

$$1\,\mathrm{kgf} = 9.807\,\mathrm{N}$$
$$1\,\mathrm{cm}^2 = 1 \times 10^{-4}\,\mathrm{m}^2$$
$$\therefore\quad 1\,\mathrm{kgf/cm}^2 = 9.807 \times 10^4\,\mathrm{N/m}^2 = 9.807 \times 10^4\,\mathrm{Pa}$$
$$\therefore\quad 120 \times 9.807 \times 10^4 = 11.8 \times 10^6\,\mathrm{Pa} = 11.8\,\mathrm{MPa}$$

△1-3

△1-3

$$1\,\mathrm{cal} = 4.1868\,\mathrm{J}$$
$$\therefore\quad 150 \times 10^3 \times 4.1868 = 628 \times 10^3\,\mathrm{J} = 628\,\mathrm{kJ}$$

△1-4

△1-4

Since those values are independent of the size, they are intensive.

それらは大きさに依存しない量であるので示強性である.

△2-1

△2-1

$$W = P(V_2 - V_1) = 0.2\,\mathrm{MPa} \times (4 - 1)\mathrm{m}^3 = 0.6\,\mathrm{MJ}$$

△2-2

△2-2

$$mgh = 1\,\mathrm{kg} \times 9.807\,\mathrm{m/s}^2 \times 1000\,\mathrm{m} = 9.807 \times 10^3\,\mathrm{J}$$
$$\therefore\quad 1\,\mathrm{kg} \times 4.185 \times 10^3\,\mathrm{J/(kg \cdot K)} \times \Delta T\,[\mathrm{K}]$$
$$\qquad = 9.807 \times 10^3\,\mathrm{J}$$
$$\therefore\quad \Delta T = 2.343\,\mathrm{K}$$

△3-1   △3-1   (3-2)

$$PV = GRT$$

$$\therefore \quad R = \frac{PV}{GT} = \frac{200 \times 10^3 \, \text{Pa} \times 2 \, \text{m}^3}{1.5 \, \text{kg} \times (25 + 273.15) \text{K}}$$

$$= 894.4 \, \text{J/(kg} \cdot \text{K)}$$

△3-2   △3-2

From Eq.3-3,   式 (3-3) より,

$$v = \frac{RT}{P}$$

As for H_2: H_2 については,

$$v = \frac{4\,124.4 \, \text{J/(kg} \cdot \text{K)} \times 293 \, \text{K}}{101.3 \times 10^3 \, \text{Pa}} = 11.93 \, \text{m}^3/\text{kg}$$

As for CO_2: CO_2 については,

$$v = \frac{188.92 \, \text{J/(kg} \cdot \text{K)} \times 293 \, \text{K}}{101.3 \times 10^3 \, \text{Pa}} = 0.546 \, \text{m}^3/\text{kg}$$

△3-3   △3-3

From Eq.3-4,   式 (3-4) より,

$$c_v = \frac{\Delta Q_v}{G(T_2 - T_1)} = \frac{22.3 \, \text{kJ}}{1 \, \text{kg} \times (303 \, \text{K} - 273 \, \text{K})}$$

$$= 0.743 \, \text{kJ/(kg} \cdot \text{K)}$$

△3-4   △3-4

From Eq.3-9,   式 (3-9) より,

$$c_p = c_v \kappa$$

Substituting this into Eq.3-8,   これを式 (3-8) に代入して,

$$c_v \kappa = c_v + R$$

$$\therefore \quad c_v (\kappa - 1) = R, \quad \text{thus} \quad c_v = \frac{R}{\kappa - 1}$$

$$\therefore \quad c_p = c_v \kappa = \frac{\kappa R}{\kappa - 1}$$

△3-5   △3-5

From Eq.3-10,   式 (3-10) より,

$$c_v = \frac{259.83 \, (\text{J/kg} \cdot \text{K})}{1.399 - 1} = 651.2 \, \text{J/(kg} \cdot \text{K)}$$

$$c_p = \frac{1.399 \times 259.83 \, (\text{J/kg} \cdot \text{K})}{1.399 - 1} = 911.0 \, \text{J/(kg} \cdot \text{K)}$$

△3-6

From Eq.3-5,

$$c_p = \frac{\Delta Q_p}{G(T_2 - T_1)} = \frac{450\,\text{kJ}}{1\,\text{kg} \times (800-300)\text{K}} = 0.9\,\text{kJ/(kg·K)}$$
$$\therefore\ R = c_p - c_v = (0.9-0.75)\,\text{kJ/(kg·K)} = 0.25\,\text{kJ/(kg·K)}$$

$$\Delta U = G\int_1^2 c_v dT = G \times c_v \int_1^2 dT = G \times c_v (T_2 - T_1)$$
$$= 1\,\text{kg} \times 0.75\,\text{kJ/(kg·K)} \times (800-300)\text{K} = 375\,\text{kJ}$$

△3-6

式 (3-5) より，

△4-1

Path A: $W = 2\,\text{MPa} \times (0.4-0.1)\text{m}^3 = 0.6\,\text{MJ}$

Path B: $W = 10\,\text{MPa} \times (0.4-0.1)\text{m}^3 = 3\,\text{MJ}$

Path C: $W = \frac{(10+2)}{2} \times (0.4-0.1)\text{m}^3 = 1.8\,\text{MJ}$

△4-2

Path A: $W = 0.1\,\text{m}^3 \times (10-2)\text{MPa} = 0.8\,\text{MJ}$

Path B: $W = 0.4\,\text{m}^3 \times (10-2)\text{MPa} = 3.2\,\text{MJ}$

Path C: $W = \frac{(0.4+0.1)}{2} \times (10-2)\text{MPa} = 2\,\text{MJ}$

△4-1

経路 A: $W = 2\,\text{MPa} \times (0.4-0.1)\text{m}^3 = 0.6\,\text{MJ}$

経路 B: $W = 10\,\text{MPa} \times (0.4-0.1)\text{m}^3 = 3\,\text{MJ}$

経路 C: $W = \frac{(10+2)}{2} \times (0.4-0.1)\text{m}^3 = 1.8\,\text{MJ}$

△4-2

経路 A: $W = 0.1\,\text{m}^3 \times (10-2)\text{MPa} = 0.8\,\text{MJ}$

経路 B: $W = 0.4\,\text{m}^3 \times (10-2)\text{MPa} = 3.2\,\text{MJ}$

経路 C: $W = \frac{(0.4+0.1)}{2} \times (10-2)\text{MPa} = 2\,\text{MJ}$

△4-3

△4-3

$$\Delta Q = \Delta U + \Delta W$$
$$\therefore\ \Delta U = \Delta Q - \Delta W$$
$$\Delta W = 0.25\,\text{MPa} \times (4-1)\text{m}^3 = 0.75\,\text{MJ}$$
$$\therefore\ \Delta U = 2 - 0.75 = 1.25\,\text{MJ}$$

△5-1

From Eqs. 5-18 and 5-19,

△5-1

式 (5-18) と式 (5-19) から，

$$PV = mN\frac{\overline{w}^2}{3} = GRT$$
$$\therefore\ \overline{w}^2 = \frac{3G}{mN}RT$$

Here,

ここで，

$$mN = G$$
$$\therefore\ \sqrt{\overline{w}^2} = \sqrt{3RT}$$

As for H$_2$ molecule,   H$_2$ 分子の平均速度：

$$R = 4124.4 \, \text{J}/(\text{kg} \cdot \text{K})$$
$$\therefore \quad \sqrt{\overline{w^2}} = \sqrt{3 \times 4124.4 \, \text{J}/(\text{kg} \cdot \text{K}) \times 300 \, \text{K}}$$
$$= 1926.6 \, \text{m/s}$$

As for CO$_2$ molecule,   CO$_2$ 分子の平均速度：

$$R = 188.92 \, \text{J}/(\text{kg} \cdot \text{K})$$
$$\therefore \quad \sqrt{\overline{w^2}} = \sqrt{3 \times 188.92 \, \text{J}/(\text{kg} \cdot \text{K}) \times 300 \, \text{K}}$$
$$= 412.34 \, \text{m/s}$$

△5-2   △5-2

From Eq. 5-43,   式（5-43）から

$$\Delta \dot{W}^G = \dot{G} \left\{ \left(\dot{h}_2 + \frac{w_2^2}{2} \right) - \left(\dot{h}_1 + \frac{w_1^2}{2} \right) + \Delta \dot{Q}_{loss} \right\}$$
$$= 0.3 \, \text{kg/s} \left\{ \left(400 \times 10^3 + \frac{(100)^2}{2} \right) - \left(250 \times 10^3 + \frac{(40)^2}{2} \right) + 7 \times 10^3 \right\} \text{J/kg}$$
$$= 48.4 \, \text{kJ/s} = 48.4 \, \text{kW}$$

△5-3   △5-3

Bernoulli's equation (Eq. 5-49):   ベルヌーイの式（5-49）より

$$P_1 + \frac{1}{2} \rho_1 w_1^2 + \rho_1 g z_1 = P_2 + \frac{1}{2} \rho_2 w_2^2 + \rho_2 g z_2$$

Here,   ここで，

$$P_1 = P_2 = \text{atmospheric pressure, and } \rho_1 = \rho_2$$
$$\therefore \quad \frac{1}{2} \left(w_2^2 - w_1^2 \right) = \frac{1}{2} w_2^2 = g(z_1 - z_2) = gh$$
$$(\because \quad w_1 \langle\langle w_2)$$
$$\therefore \quad w_2 = \sqrt{2gh}$$

△6-1   △6-1

The entropy loss of heat source: 熱源が失ったエントロピー

$$\frac{\Delta Q}{T_{source}} = \frac{120 \, \text{kJ}}{1\,000 \, \text{K}} = 0.12 \, \text{kJ/K}$$

The entropy gain of working fluid: 作動流体が得たエントロピー

$$\frac{\Delta Q}{T_{fluid}} = \frac{120 \, \text{kJ}}{300 \, \text{K}} = 0.4 \, \text{kJ/K}$$

The total entropy change:　　　　　　　　　　　全体のエントロピー変化

$$\Delta S = \frac{\Delta Q}{T_{fluid}} - \frac{\Delta Q}{T_{source}} = 0.4 - 0.12 = 0.28 \, \text{kJ/K}$$

Then, the total entropy increases.　　　　　　したがって，全体のエントロピーは増加する．

△6-2　　　　　　　　　　　　　　　　　　　△6-2

From Eq. 6-48,　　　　　　　　　　　　　　式（6-48）より，

$$dH = dU + PdV + VdP$$

Substitute this for Eq. 6-47,　　　　　　　　この式を式（6-47）に代入して，

$$TdS = dU + PdV = (dH - PdV - VdP) + PdV$$
$$= dH - VdP$$

△6-3　　　　　　　　　　　　　　　　　　　△6-3

$$G = H - TS = (U + PV) - TS$$
$$= (U - TS) + PV = F + PV$$

△7-1　　　　　　　　　　　　　　　　　　　△7-1

The amount of heat is represented by the area under the process curve on a T-S chart.　熱量は T-S 線図において状態変化を示す線の下部の面積で表される．

$$Q_{12} = \frac{(800 - 300)\text{K}}{2} \times (120 - 100) \, \text{kJ/K}$$
$$= 5\,000 \, \text{kJ} = 5 \, \text{MJ}$$

△7-2　　　　　　　　　　　　　　　　　　　△7-2

$$dS_w = \frac{dQ}{T_w} = \frac{c_w dT_w}{T_w}$$
$$\therefore \quad \Delta S_w = \int_{T_1}^{T_2} \frac{c_w dT_w}{T_w} = c_w \ln \frac{T_2}{T_1}$$
$$\therefore \quad \frac{\Delta S_w}{c_w} = \ln \frac{T_2}{T_1}$$
$$\therefore \quad T_2 = T_1 \exp\left(\frac{\Delta S_w}{c_w}\right) = 278.15 \, \text{K} \times \exp\left(\frac{1.2 \, \text{kJ/K}}{4.19 \, \text{kJ/(kg·K)}}\right)$$
$$= 371 \, \text{K} = 97.5°\text{C}$$

△8-1　　　　　　　　　　　　　　　　　　　△8-1

(1)　　　　　　　　　　　　　　　　　　　　（1）

$$P_1 V_1 = P_2 V_2$$
$$\therefore \quad P_2 = \frac{P_1 V_1}{V_2} = \frac{5 \, \text{MPa} \times 0.2 \, \text{m}^3}{4 \, \text{m}^3} = 0.25 \, \text{MPa}$$

(2)

(2)

$$GRT = P_1V_1 = P_2V_2$$
$$W_{12} = L_{12} = GRT \ln \frac{P_1}{P_2} = P_1V_1 \ln \frac{P_1}{P_2}$$
$$= 5\,\text{MPa} \times 0.2\,\text{m}^3 \times \ln \frac{5\,\text{MPa}}{0.25\,\text{MPa}} = 3.0\,\text{MJ}$$

(3)

(3)

$$\Delta S = GR \ln \frac{P_1}{P_2} = \frac{P_1V_1}{T} \ln \frac{P_1}{P_2}$$
$$= \frac{5 \times 10^6\,\text{Pa} \times 0.2\,\text{m}^3}{300\,\text{K}} \ln \frac{5\,\text{MPa}}{0.25\,\text{MPa}} = 10.0\,\text{kJ/K}$$

△8-2

△8-2

(1)

(1)

$$V_1 = \frac{GRT}{P} = \frac{3\,\text{kg} \times 0.2871 \times 10^3\,\text{J/(kg·K)} \times 300\,\text{K}}{0.1 \times 10^6\,\text{Pa}} = 2.584\,\text{m}^3$$
$$\therefore\quad V_2 = \frac{T_2}{T_1}V_1 = \frac{400\,\text{K}}{300\,\text{K}} \times 2.584\,\text{m}^3 = 3.445\,\text{m}^3$$

(2)

(2)

$$W_{12} = P(V_2 - V_1) = 0.1\,\text{MPa} \times (3.445 - 2.584)\,\text{m}^3 = 86.1\,\text{kJ}$$

(3)

(3)

$$\Delta S = Gc_p \ln \frac{T_2}{T_1} = 3\,\text{kg} \times 1.005\,\text{kJ/(kg·K)} \times \ln \frac{400\,\text{K}}{300\,\text{K}} = 867.4\,\text{J/K}$$

△8-3

△8-3

(1)

(1)

$$P_1 = \frac{GRT}{V} = \frac{3\,\text{kg} \times 0.2871 \times 10^3\,\text{J/(kg·K)} \times 250\,\text{K}}{0.1\,\text{m}^3} = 2.153\,\text{MPa}$$
$$\therefore\quad P_2 = \frac{T_2}{T_1}P_1 = \frac{450\,\text{K}}{250\,\text{K}} \times 2.153\,\text{MPa} = 3.875\,\text{MPa}$$

(2)

(2)

$$Q_{12} = Gc_v(T_2 - T_1) = 3\,\text{kg} \times 0.716\,\text{kJ(kg·K)} \times (450 - 250)\,\text{K} = 429.6\,\text{kJ}$$

(3)

(3)

$$\Delta S = Gc_v \ln \frac{T_2}{T_1} = 3\,\text{kg} \times 0.716\,\text{kJ/(kg·K)} \times \ln \frac{450\,\text{K}}{250\,\text{K}} = 1.263\,\text{kJ/K}$$

△8-4
(1)

△8-4
(1)

$$\kappa = \frac{c_p}{c_v} = \frac{c_p}{c_p - R} = \frac{1.005\,\text{kJ}/(\text{kg}\cdot\text{K})}{(1.005 - 0.2871)\,\text{kJ}/(\text{kg}\cdot\text{K})} = 1.40$$

$$\therefore \quad P_2 = P_1\left(\frac{V_1}{V_2}\right)^{\kappa} = 0.2\,\text{MPa} \times \left(\frac{1\,\text{m}^3}{0.2\,\text{m}^3}\right)^{1.40} = 1.9\,\text{MPa}$$

(2)

(2)

$$T_2 = T_1\left(\frac{V_1}{V_2}\right)^{\kappa-1} = 300\,\text{K} \times \left(\frac{1\,\text{m}^3}{0.2\,\text{m}^3}\right)^{0.4} = 571\,\text{K}$$

(3)

(3)

$$c_v = c_p - R = 0.718\,\text{kJ}/(\text{kg}\cdot\text{K})$$

$$G = \frac{P_1 V_1}{R T_1} = \frac{0.2 \times 10^6\,\text{Pa} \times 1\,\text{m}^3}{0.2871 \times 10^3\,\text{J}/(\text{kg}\cdot\text{K}) \times 300\,\text{K}} = 2.322\,\text{kg}$$

$$\therefore \quad W_{12} = -G c_v (T_2 - T_1) = -2.322\,\text{kg} \times 0.718\,\text{kJ}/(\text{kg}\cdot\text{K}) \times (571 - 300)\,\text{K}$$

$$= -451.8\,\text{kJ}$$

△8-5

△8-5

$$P_1 V_1^n = P_2 V_2^n$$

$$\therefore \quad \frac{P_1}{P_2} = \left(\frac{V_2}{V_1}\right)^n$$

$$\ln\left(\frac{P_1}{P_2}\right) = n \ln\left(\frac{V_2}{V_1}\right)$$

$$\therefore \quad n = \ln\left(\frac{P_1}{P_2}\right) \Big/ \ln\left(\frac{V_2}{V_1}\right) = \ln\left(\frac{0.1\,\text{MPa}}{15\,\text{MPa}}\right) \Big/ \ln\left(\frac{0.078\,\text{m}^3}{5\,\text{m}^3}\right) = 1.20$$

△9-1
The work estimated theoretically

△9-1
理論的に求めた仕事

$$W_{T12} = GR(T_2 - T_1) = 2\,\text{kg} \times 0.1889\,\text{kJ}/(\text{kg}\cdot\text{K}) \times (500 - 300)\,\text{K}$$

$$= 75.56\,\text{kJ}$$

$$\therefore \quad \text{Energy Loss} = 75.56 - 70 = 5.56\,\text{kJ}$$

△10-1
(1) y_i : mole fraction of species i

△10-1
(1) y_i : 化学種 i のモル分率

Volume fraction is equal to mole fraction

$$\therefore \quad y_{O2} = 0.21, \quad y_{N2} = 0.79$$

M_m : The average molar mass of the mixture

$$M_m = y_{O2} M_{O2} + y_{N2} M_{N2} = 0.21 \times 32.00\,\text{kg}/\text{kmol} + 0.79 \times 28.02\,\text{kg}/\text{kmol}$$

$$= 28.86\,\text{kg}/\text{kmol}$$

(2) m_i : mass fraction of species i   (2) m_i : 化学種 i の質量分率

$$m_{O2} = y_{O2} \frac{M_{O2}}{M_m} = 0.21 \times \frac{32.00 \text{ kg/kmol}}{28.86 \text{ kg/kmol}} = 0.233$$

$$m_{N2} = y_{N2} \frac{M_{N2}}{M_m} = 0.79 \times \frac{28.02 \text{ kg/kmol}}{28.86 \text{ kg/kmol}} = 0.767$$

(3) R_m : average gas constant   (3) R_m : 混合ガスのガス定数

$$R_m = m_{O2} R_{O2} + m_{N2} R_{N2}$$
$$= 0.233 \times 0.2598 \text{ kJ/(kg·K)} + 0.767 \times 0.2968 \text{ kJ/(kg·K)} = 0.2882 \text{ kJ/(kg·K)}$$

△10-2   △10-2

$$P_a V_d = (P_1 - P_2) \frac{V}{\kappa}$$

$$\therefore \quad V_d = \frac{P_1 - P_2}{P_a} \frac{V}{\kappa} = \frac{(1 \times 10^6 - 0.5 \times 10^6) \text{Pa}}{0.1 \times 10^6 \text{ Pa}} \times \frac{10 \text{ m}^3}{1.40} = 35.7 \text{ m}^3$$

$$\frac{T_1}{P_1{}^{\frac{\kappa-1}{\kappa}}} = \frac{T_2}{P_2{}^{\frac{\kappa-1}{\kappa}}} \quad \therefore T_2 = T_1 \left(\frac{P_2}{P_1} \right)^{\frac{\kappa-1}{\kappa}} = 300 \text{ K} \times \left(\frac{0.5 \text{ MPa}}{1 \text{ MPa}} \right)^{\frac{0.4}{1.4}} = 246 \text{ K}$$

$$\therefore \quad G_d = \left(\frac{P_1}{T_1} - \frac{P_2}{T_2} \right) \frac{V}{R} = \left(\frac{1 \times 10^6 \text{ Pa}}{300 \text{ K}} - \frac{0.5 \times 10^4 \text{ Pa}}{246 \text{ K}} \right) \times \left(\frac{10 \text{ m}^3}{0.287 \times 10^3 \text{ J/(kg·K)}} \right) = 45.3 \text{ kg}$$

△10-3   △10-3

$$T_2 = \frac{\kappa T_1 T_f}{T_1 + (\kappa T_f - T_1) \frac{P_1}{P_2}} = \frac{1.40 \times 300 \text{ K} \times 280 \text{ K}}{300 \text{ K} + (1.40 \times 280 \text{ K} - 300 \text{ K}) \frac{0.2 \text{ MPa}}{10 \text{ MPa}}} = 390 \text{ K}$$

$$G_f = \frac{P_2 V}{R T_2} - \frac{P_1 V}{R T_1} = \frac{10 \times 10^6 \text{ Pa} \times 3 \text{ m}^3}{0.287 \times 10^3 \text{ J/(kg·K)} \times 390 \text{ K}} - \frac{0.2 \times 10^6 \text{ Pa} \times 3 \text{ m}^3}{0.287 \times 10^3 \text{ J/(kg·K)} \times 300 \text{ K}}$$
$$= 261 \text{ kg}$$

△11-1   △11-1

$$C = \sqrt{\kappa R T} = \sqrt{1.40 \times 0.287 \times 10^3 \text{ kJ/(kg·K)} \times (273.15 + 20) \text{K}} = 343.2 \text{ m/s}$$

$$M = \frac{300 \text{ m/s}}{343.2 \text{ m/s}} = 0.874$$

△11-2   △11-2

$$w_2 = \sqrt{2 \frac{\kappa}{\kappa-1} P_1 v_1 \left\{ 1 - \left(\frac{P_2}{P_1} \right)^{\frac{\kappa-1}{\kappa}} \right\}} = \sqrt{2 \frac{\kappa}{\kappa-1} R T_1 \left\{ 1 - \left(\frac{P_2}{P_1} \right)^{\frac{\kappa-1}{\kappa}} \right\}}$$

$$= \sqrt{2 \times \frac{1.40}{0.4} \times 0.287 \times 10^3 \text{ kJ/(kg·K)} \times 300 \text{ K} \left\{ 1 - \left(\frac{0.1 \text{ MPa}}{0.2 \text{ MPa}} \right)^{\frac{0.4}{1.4}} \right\}} = 329.1 \text{ m/s}$$

△11-3

△11-3

$$\eta_n = \frac{w_2'^2}{w_2^2} = \left(\frac{320\,\mathrm{m/s}}{329.1\,\mathrm{m/s}}\right)^2 = 0.945$$

△11-4

△11-4

$$P_c = P_1\left(\frac{2}{\kappa+1}\right)^{\frac{\kappa}{\kappa-1}} = 0.5\,\mathrm{MPa}\times\left(\frac{2}{2.4}\right)^{\frac{1.4}{0.4}} = 0.264\,\mathrm{MPa}$$

(1) Therefore, if the back pressure P_2 is 0.3 MPa, the flow is not choked.

(1) したがって，背圧 P_2 が 0.3 MPa の場合には流れは閉塞していない．

$$\therefore\ \dot{G} = A_2\sqrt{2\frac{\kappa}{\kappa-1}\frac{P_1}{v_1}\left\{\left(\frac{P_2}{P_1}\right)^{\frac{2}{\kappa}}-\left(\frac{P_2}{P_1}\right)^{\frac{\kappa+1}{\kappa}}\right\}} = A_2\sqrt{2\frac{\kappa}{\kappa-1}\frac{P_1^2}{RT_1}\left\{\left(\frac{P_2}{P_1}\right)^{\frac{2}{\kappa}}-\left(\frac{P_2}{P_1}\right)^{\frac{\kappa+1}{\kappa}}\right\}}$$

$$= 400\times10^{-6}\sqrt{2\times\frac{1.4}{0.4}\times\frac{\left(0.5\times10^6\,\mathrm{Pa}\right)^2}{0.287\times10^3\,\mathrm{J/(kg\cdot K)}\times303.15\,\mathrm{K}}\left\{\left(\frac{0.3\,\mathrm{MPa}}{0.5\,\mathrm{MPa}}\right)^{\frac{2}{1.4}}-\left(\frac{0.3\,\mathrm{MPa}}{0.5\,\mathrm{MPa}}\right)^{\frac{2.4}{1.4}}\right\}}$$

$$= 0.459\,\mathrm{kg/s}$$

(2) If the back pressure P_2 is 0.1 MPa, the flow is choked.

(2) 背圧 P_2 が 0.1 MPa の場合には流れは閉塞する．

$$\therefore\ \dot{G} = A_2\sqrt{2\frac{\kappa}{\kappa+1}\left(\frac{2}{\kappa+1}\right)^{\frac{2}{\kappa-1}}\frac{P_1}{v_1}} = A_2\sqrt{2\frac{\kappa}{\kappa+1}\left(\frac{2}{\kappa+1}\right)^{\frac{2}{\kappa-1}}\frac{P_1^2}{RT_1}}$$

$$= 400\times10^{-6}\times\sqrt{2\times\frac{1.4}{2.4}\times\left(\frac{2}{2.4}\right)^{\frac{2}{0.4}}\times\frac{\left(0.5\times10^6\,\mathrm{Pa}\right)^2}{0.287\times10^3\,\mathrm{J/(kg\cdot K)}\times303.15\,\mathrm{K}}}$$

$$= 0.464\,\mathrm{kg/s}$$

△12-1

△12-1

$$W_c = Q_H - Q_L = (500-320)\,\mathrm{kJ} = 180\,\mathrm{kJ}$$
$$\eta = \frac{W_c}{Q_H} = \frac{180\,\mathrm{kJ}}{500\,\mathrm{kJ}} = 0.36 = 36\%$$

△12-2

△12-2

$$[\mathrm{COP}]_r = \frac{Q_L}{W_c}$$
$$\therefore\ Q_L = [\mathrm{COP}]_r\times W_c = 2.50\times5\,\mathrm{kJ} = 12.5\,\mathrm{kJ}$$

△12-3

△12-3

(1)

(1)

$$Q_H = T_H\Delta S = 1\,000\,\mathrm{K}\times300\,\mathrm{J/K} = 300\,\mathrm{kJ}$$

(2) Since this cycle is reversible, the received entropy is equal to the rejected entropy.

(2) このサイクルは可逆であるから受け取ったエントロピーと放出したエントロピーは等しい.

$$\therefore \frac{Q_H}{T_H} = \frac{Q_L}{T_L}$$

$$T_L = \frac{Q_L}{Q_H} T_H = \frac{120\,\text{kJ}}{300\,\text{kJ}} \times 1000\,\text{K} = 400\,\text{K}$$

(3)

(3)

$$\eta = 1 - \frac{Q_L}{Q_H} = 1 - \frac{120\,\text{kJ}}{300\,\text{kJ}} = 0.60 = 60.0\%$$

△13-1
(1)

△13-1
(1)

$$\eta = 1 - \frac{T_H}{T_L} = 1 - \frac{300\,\text{K}}{1\,000\,\text{K}} = 0.7 = 70\%$$

(2)

(2)

$$\eta = \frac{G \cdot w_c}{G \cdot q_H} = \frac{w_c}{q_H}$$
$$\therefore w_c = \eta \cdot q_H = 0.7 \times 300\,\text{kJ/kg} = 210\,\text{kJ/kg}$$

(3)

(3)

$$\eta = 1 - \frac{q_L}{q_H}$$
$$\therefore q_L = q_H(1-\eta) = 300\,\text{kJ/kg} \times (1-0.7) = 90\,\text{kJ/kg}$$

△13-2
(1)

△13-2
(1)

$$\eta = 1 - \frac{T_L}{T_H} = 1 - \frac{400\,\text{K}}{1\,000\,\text{K}} = 0.6$$
$$\eta = \frac{W_c}{Q_H} = \frac{Q_H - Q_L}{Q_L}$$
$$\therefore \eta Q_H = Q_H - Q_L = 0.48\,\text{kJ}$$
$$\therefore Q_H = \frac{W_c}{\eta} = \frac{0.48\,\text{kJ}}{0.6} = 0.8\,\text{kJ}$$
$$\therefore Q_L = Q_H - W_c = 0.8\,\text{kJ} - 0.48\,\text{kJ} = 0.32\,\text{kJ}$$

(2)

(2)

$$\Delta S = \frac{Q_L}{T_L} - \frac{Q_H}{T_H} = \frac{0.32\,\text{kJ}}{400\,\text{K}} - \frac{0.8\,\text{kJ}}{1\,000\,\text{K}} = 0\,\text{kJ/K}$$

△ 13-3 　　　　　　　　　　　　　　　　　△ 13-3

$$\Delta s_{1\text{-}2} = \frac{q_a}{T_a}, \qquad \Delta s_{3\text{-}4} = -\frac{q_b}{T_b}$$

Since this cycle is reversible, 　　　　　　　このサイクルは可逆サイクルだから

$$\Delta s_{1\text{-}2} + \Delta s_{3\text{-}4} = 0, \qquad (\because \quad \Delta s_{2\text{-}3} = \Delta s_{4\text{-}1} = 0)$$

$$\therefore \quad \frac{q_a}{T_a} - \frac{q_b}{T_b} = 0$$

$$\therefore \quad T_b = T_a \frac{q_b}{q_a} = 1\,000\,\text{K} \times \frac{16\,\text{kJ/kg}}{40\,\text{kJ/kg}} = 400\,\text{K}$$

$$\therefore \quad \Delta s_{3\text{-}4} = \frac{q_b}{T_b} = \frac{16\,\text{kJ/kg}}{400\,\text{K}} = 40\,\text{J/(kg·K)}$$

On the other hand, 　　　　　　　　　　　一方，

$$\Delta s_{3\text{-}4} = R\,\ln\frac{P_4}{P_3}, \quad \therefore \quad P_4 = P_3 \exp\!\left(\frac{\Delta s_{3\text{-}4}}{R}\right) = 0.050\,\text{MPa} \times \exp\!\left(\frac{40\,\text{J/(kg·K)}}{0.287\times10^3\,\text{J/(kg·K)}}\right) = 0.057\,\text{MPa}$$

△ 14-1 　　　　　　　　　　　　　　　　　△ 14-1

$$\eta_{Otto} = 1 - \frac{1}{\varepsilon_{Otto}^{\kappa-1}} = 1 - \frac{1}{(9.9)^{0.4}} = 0.60 = 60\%$$

△ 14-2 　　　　　　　　　　　　　　　　　△ 14-2

$$\eta_{Diesel} = 1 - \frac{1}{\varepsilon_{Diesel}^{\kappa-1}} \cdot \frac{\sigma^\kappa - 1}{\kappa(\sigma-1)} = 1 - \frac{1}{(18.0)^{0.4}} \times \frac{(4.0)^{1.4}-1}{1.4\times(4.0-1)} = 0.553 = 55.3\%$$

△ 14-3 　　　　　　　　　　　　　　　　　△ 14-3

$$\eta_{Sabathe} = 1 - \frac{1}{\varepsilon_{Sabathe}^{\kappa-1}} \cdot \frac{\xi\sigma^\kappa - 1}{(\xi-1)+\xi\kappa(\sigma-1)} = 1 - \frac{1}{(18.0)^{0.4}} \times \frac{5.0\times(2.0)^{1.4}-1}{(5.0-1)+5.0\times1.4\times(2.0-1)} = 0.651 = 65.1\%$$

△ 14-4 　　　　　　　　　　　　　　　　　△ 14-4

$$\eta_{Brayton} = 1 - \frac{1}{\lambda^{\frac{\kappa-1}{\kappa}}} = 1 - \frac{1}{(12.0)^{\frac{0.4}{1.4}}} = 0.508 = 50.8\%$$

△ 15-1 　　　　　　　　　　　　　　　　　△ 15-1
(1) 　　　　　　　　　　　　　　　　　　　(1)

$$P = \frac{RT}{v} = \frac{0.2870\,\text{kJ/(kg·K)} \times 180\,\text{K}}{0.00285\,\text{m}^3/\text{kg}} = 18\,126\,\text{kPa}$$

(2)   (2)

$$P = \frac{RT}{v-b} - \frac{a}{v^2} = \frac{0.2870\,\mathrm{kJ/(kg\cdot K)} \times 180\,\mathrm{K}}{(0.00285 - 0.00126)\,\mathrm{m^3/kg}} - \frac{0.162\,\mathrm{m^6\,kPa/kg}}{\left(0.00285\,\mathrm{m^3/kg}\right)^2}$$
$$= 12\,546\,\mathrm{kPa}$$

△15-2   △15-2

$$v = (1-x)v' + xv'' = (1-0.5) \times 0.001061\,\mathrm{m^3/kg} + 0.5 \times 0.8857\,\mathrm{m^3/kg} = 0.4434\,\mathrm{m^3/kg}$$
$$s = (1-x)s' + xs'' = (1-0.5) \times 1\,530\,\mathrm{kJ/(kg\cdot K)} + 0.5 \times 7\,127\,\mathrm{kJ/(kg\cdot K)} = 4\,329\,\mathrm{kJ/(kg\cdot K)}$$

△15-3   △15-3

$$\eta_{Rankine} = \frac{(h_1-h_2)-(h_4-h_3)}{(h_1-h_3)-(h_4-h_3)} = \frac{\left(3\,365\,\mathrm{kJ/kg} - 2\,012\,\mathrm{kJ/kg}\right) - \left(138\,\mathrm{kJ/kg} - 138\,\mathrm{kJ/kg}\right)}{\left(3\,365\,\mathrm{kJ/kg} - 138\,\mathrm{kJ/kg}\right) - \left(138\,\mathrm{kJ/kg} - 138\,\mathrm{kJ/kg}\right)} = 0.419 = 41.9\%$$

Index
索　引

English-Japanese ／英-和

A

M

N

O

P

Q

R

S

Index
索　引

Japanese-English ／和-英

著 者 略 歴

新井　雅隆（あらい・まさたか）
1977 年　東北大学大学院博士課程修了
　　　　Dr. Eng'g, Graduate School of Tohoku Univ.
1977 年　広島大学工学部助手
　　　　Reserch Associate, Faculty of Eng'g, Hiroshima Univ.
1994 年　群馬大学工学部教授
　　　　Professor, Faculty of Eng'g, Gunma Univ.
2007 年　群馬大学大学院工学研究科教授
　　　　Professor, Graduate School of Eng'g, Gunma Univ.

古畑　朋彦（ふるはた・ともひこ）
1994 年　東北大学大学院博士後期課程修了
　　　　Dr. Eng'g, Graduate School of Tohoku Univ.
1994 年　東北大学工学部助手
　　　　Research Associate, Faculty of Eng'g, Tohoku Univ.
1997 年　名古屋大学高温エネルギー変換研究センター助手
　　　　Research Associate, Research Center for Advanced Energy Conversion,
　　　　Nagoya Univ.
2002 年　名古屋大学工学部講師
　　　　Lecturer, Faculty of Eng'g, Nagoya Univ.
2004 年　群馬大学工学部助教授
　　　　Associate Professor, Faculty of Eng'g, Gunma Univ.
2007 年　群馬大学大学院工学研究科准教授
　　　　Associate Professor, Graduate School of Eng'g, Gunma Univ.

英和対照「工学基礎テキスト」シリーズ
熱力学　　　　　　　　　　　　　　　　© 新井雅隆・古畑朋彦　2008

2008 年 12 月 20 日　第 1 版第 1 刷発行　　【本書の無断転載を禁ず】

著　　　者　新井雅隆・古畑朋彦

発 行 者　森北博巳

発 行 所　森北出版株式会社

　　　　　東京都千代田区富士見 1-4-11（〒102-0071）
　　　　　電話 03-3265-8341／FAX 03-3264-8709
　　　　　http://www.morikita.co.jp/
　　　　　日本書籍出版協会・自然科学書協会・工学書協会　会員
　　　　　JCLS ＜(株)日本著作出版権管理システム委託出版物＞

落丁・乱丁本はお取替えいたします　　　　印刷／双文社印刷・製本／協栄製本

Printed in Japan／ISBN 978-4-627-63021-5

英和対照「工学基礎テキスト」シリーズ

熱力学 ［POD 版］

2022 年 11 月 25 日発行

著者　　　新井雅隆・古畑朋彦

印刷　　　大日本印刷株式会社
製本　　　大日本印刷株式会社

発行者　　森北博巳
発行所　　森北出版株式会社
　　　　　〒102-0071　東京都千代田区富士見 1-4-11
　　　　　03-3265-8342（営業・宣伝マネジメント部）
　　　　　https://www.morikita.co.jp/